3000 800059 80769

St. Louis Community College

Meramec Library
St. Louis Community College
11333 Big Bend Blvd.
Kirkwood, MO 63122-5799
314-984-7797

A.A. Librerie Universitaire
di Bologna
V.LE A.L.V.

ENERGY
In The 21st Century

St. Louis Community College
at Meramec
LIBRARY

ENERGY

In The 21st Century

JOHN R FANCHI

Colorado School of Mines, USA

World Scientific

NEW JERSEY • LONDON • SINGAPORE • BEIJING • SHANGHAI • HONG KONG • TAIPEI • CHENNAI

Published by

World Scientific Publishing Co. Pte. Ltd.

5 Toh Tuck Link, Singapore 596224

USA office: 27 Warren Street, Suite 401-402, Hackensack, NJ 07601

UK office: 57 Shelton Street, Covent Garden, London WC2H 9HE

British Library Cataloguing-in-Publication Data
A catalogue record for this book is available from the British Library.

ENERGY IN THE 21ST CENTURY

Copyright © 2005 by World Scientific Publishing Co. Pte. Ltd.

All rights reserved. This book, or parts thereof, may not be reproduced in any form or by any means, electronic or mechanical, including photocopying, recording or any information storage and retrieval system now known or to be invented, without written permission from the Publisher.

For photocopying of material in this volume, please pay a copying fee through the Copyright Clearance Center, Inc., 222 Rosewood Drive, Danvers, MA 01923, USA. In this case permission to photocopy is not required from the publisher.

ISBN 981-256-185-4
ISBN 981-256-195-1 (pbk)

Editor: Tjan Kwang Wei

Printed in Singapore.

This book is dedicated to those who support
a sustainable energy future.

PREFACE

My interest in energy began in the 1970's when I obtained degrees in physics from the Universities of Denver (B.S.), Mississippi (M.S.), and Houston (Ph.D.). I did some work in geothermal storage of solar energy as a post-doc in 1978, and then spent many years in the energy industry helping develop oil and gas reservoirs. I became a full time academic in 1998 when I joined the faculty of the Colorado School of Mines as a professor of petroleum engineering.

In the transition from industry to academia, I wanted to find out how long a college graduate today could expect to continue a career in the extraction of fossil fuels. After studying several forecasts of energy production, I was convinced that fossil fuels would continue to be an important part of the energy mix while other energy sources would increase in importance. To help prepare students to function as energy professionals, I developed an energy course at the Colorado School of Mines and published the textbook **Energy: Technology and Directions for the Future** (Elsevier – Academic Press, Boston, 2004).

I realized as I was developing the energy course that much of the material in the textbook is suitable for a general audience. This book, **Energy in the 21st Century**, is a non-technical version of **Energy: Technology and Directions for the Future**. **Energy in the 21st Century** was written to give the concerned citizen enough information about energy to make informed decisions. Readers who would like more detailed information or a more complete list of references should consult the textbook **Energy: Technology and Directions for the Future**.

I want to thank my students and guest speakers for their comments during the preparation of my energy course. Tony Fanchi helped prepare many of the figures in the book, and Kathy Fanchi was instrumental in the preparation and production of the book. Even though there are many more topics that could be discussed, the material in **Energy in the 21ˢᵗ Century** should expose you to a broad range of energy types and help you develop an appreciation of the role that each energy type may play in the future.

John R. Fanchi
January 2005

CONTENTS

Chapter 1

A BRIEF HISTORY OF ENERGY CONSUMPTION

We all make decisions about energy. We decide how much electricity we will use to heat or cool our homes. We decide how far we will drive every day and the type of vehicle we will drive. Those of us in democracies choose leaders who create budgets that can support new energy initiatives or maintain a military capable of defending energy supply lines. Each of these decisions and many others impact the global consumption of energy and the demand for available natural resources. The purpose of this book is to give you the information you need to help you make informed decisions.

Oil has been the fuel of choice to meet our energy needs. Many experts believe that the supply of oil will reach a peak in the first quarter of the 21st century – sometime between now and 2025 – and will begin to decline. The decline in produced oil will occur despite an increasing demand for energy. The combination of increased demand and reduced supply will lead to a significant increase in the price of oil. According to this scenario, other sources of energy will begin to replace the increasingly expensive oil. The dominance of oil in the current energy mix will not continue.

Ever since the first oil crisis in 1973, alarmists have made dire predictions in the media that the price of oil will increase with virtually no limit. Will their predictions come true, or have they neglected market forces that constrain the price of oil and other fossil fuels? We will examine how society can move from a dependence on oil to energy independence as we consider answers to the question: how should society make the transition from fossil fuels to energy sources that will serve humanity indefinitely?

The choices we make today will affect generations to come. What kind of future do we want to prepare for them? What kind of future is possible? We can make the best decisions by being aware of our options and the consequences of our choices. In this book, we consider the location, quantity and accessibility of energy sources. We discuss ways to distribute available energy, and examine how our choices will affect the economy, society, and the environment. Our understanding of each of these issues will help us on our journey to energy independence. We begin by finding out where we are. We begin with a review of our history of energy consumption.

HISTORICAL PER CAPITA ENERGY CONSUMPTION

The history of energy consumption shows how important energy is to the quality of life for each of us. Societies have depended on different types of energy in the past, and societies have been forced to change from one energy type to another. Global energy consumption can be put in perspective by considering the amount of energy consumed by individuals.

E. Cook [1971] provided estimates of daily human energy consumption at six different periods of societal development. Cook's estimates are given in Table 1-1. The table shows that personal energy consumption has increased as society has evolved.

In ancient times, energy was consumed in the form of food. Cook assumed the only source of energy consumed by a person living during the period labeled "Primitive" was food. Energy is essential for life, and food was the first source of energy. According to Cook, humans require approximately 2000 kilocalories (about eight megajoules) of food per day. One food Calorie is equal to one kilocalorie. One calorie is the amount of energy required to raise the temperature of one gram of water one degree centigrade. A change in temperature of one degree Centigrade is equal to a change in temperature of 1.8 degrees Fahrenheit. Cook's energy estimate was for an East African about one million years ago.

Table 1-1. Historical Energy Consumption [Cook, 1971]					
Period	Daily per capita Consumption (1000 kcal)				
	Food	H & C*	I & A**	Transportation	Total
Primitive	2				2
Hunting	3	2			5
Primitive Agricultural	4	4	4		12
Advanced Agricultural	6	12	7	1	26
Industrial	7	32	24	14	77
Technological	10	66	91	63	230
* H & C = Home and Commerce ** I & A = Industry and Agriculture					

The ability to control fire during the Hunting period let people use wood to heat and cook. Fire provided light at night and could illuminate caves. Firewood was the first source of energy for consumption in a residential setting. Cook's estimate of the daily per capita energy consumption for Europeans about 100,000 years ago was 5,000 kilocalories (about 21 megajoules in metric units).

The Primitive Agricultural period was characterized by the domestication of animals. Humans were able to use animals to help them grow crops and cultivate their fields. The ability to grow more food than you needed became the impetus for creating an agricultural industry. Cook's estimate of the daily per capita energy consumption for people in the Fertile Crescent circa 5000 B.C.E. (Before Common Era) was 12,000 kilocalories (about 50 megajoules). Humans continue to use animals to perform work (Figure 1-1).

Figure 1-1. Animal Labor in Ahmedabad, India

More energy was consumed during the Advanced Agricultural period when people learned to use coal, and built machines to harvest the wind and water. By the early Renaissance, people were using wind to push sailing ships, water to drive mills, and wood and coal for generating heat. An example of a traditional windmill is shown in Figure 1-2. Transportation became a significant component of energy consumption by humans. Cook's estimate of the daily per capita energy consumption for people in northwestern Europe circa 1400 C.E. (Common Era) was 26,000 kilo-calories (about 109 megajoules).

Figure 1-2. Traditional Windmill in Holland

The steam engine ushered in the Industrial period. It provided a means of transforming heat energy to mechanical energy. Wood was the first source of energy for generating steam in steam engines. Coal, a fossil fuel, eventually replaced wood and hay as the primary energy source in industrialized nations. Coal was easier to store and transport than wood and hay, which are bulky and awkward. Coal was useful as a fuel source for large vehicles, such as trains and ships, but of limited use for personal transportation. Oil, another fossil fuel, was a liquid and contained about the same amount of energy per unit mass as coal. Oil could flow through pipelines and tanks. People just needed a machine to convert the energy in oil to a more useful form. Cook's estimate of the daily per capita energy consumption for people in England circa 1875 C.E. was 77,000 kilocalories (about 322 megajoules).

The modern Technological period is associated with the development of internal combustion engines, and applications of electricity. Internal combustion engines use oil and can vary widely in size. The internal combustion engine could be scaled to fit on a wagon and create "horse-less carriages." The transportation system in use today evolved as a result of the development of internal combustion engines. Electricity, by contrast, is generated from primary energy sources such as fossil fuels. Electricity generation and distribution systems made the widespread use of electric motors and electric lights possible. One advantage of electricity as an energy source is that it can be transported easily, but electricity is difficult to store. Cook's estimate of the daily per capita energy consumption for people in the United States circa 1970 C.E. was 230,000 kilocalories (about 962 megajoules).

Point to Ponder: What does an energy unit mean to me?
To get an idea of the meaning of an energy unit such as

kilocalorie or megajoule, we note that a 1200 Watt hair dryer uses approximately one megajoule of energy in 15 minutes. A 100 Watt light bulb will use one megajoule of energy in about three hours. [Fanchi, 2004, Exercise 1-3]

ENERGY CONSUMPTION AND THE QUALITY OF LIFE

The existing per capita energy consumption discussed above gives us an idea of how much energy each of us uses today, but it does not tell us how much energy each of us should use. One way to estimate the amount of energy each person should use is to examine the relationship between energy consumption and quality of life.

Quality of life is a subjective concept that can be quantified in several ways. The United Nations calculates a quantity called the Human Development Index (HDI) to provide a quantitative measure of the quality of life. The HDI measures human development in a country using three different categories: life expectancy, education, and Gross Domestic Product (GDP). Gross Domestic Product accounts for the total output of goods and services from a nation and is a measure of the economic growth of the nation. The HDI is a fraction that varies from zero to one. A value of HDI that approaches zero is considered a relatively low quality of life, while a value of HDI that approaches one is considered a high quality of life.

A plot of HDI versus per capita electricity consumption for all nations with a population of at least one million people is shown in Figure 1-3. Per capita electricity consumption is the total amount of electricity consumed by the nation divided by the population of the nation. It represents an average amount of electricity consumed by each individual in the nation. The calculation of per capita electricity consumption establishes a common basis for comparing the consumption of electricity between nations with large populations and nations with small populations. The HDI data are 1999 data from the 2001 United Nations Human Development Report [UNDP, 2001], and annual per capita electricity consumption data are 1999 data reported by the Energy Information Administration of the United States Department of Energy [EIA Table 6.2, 2002].

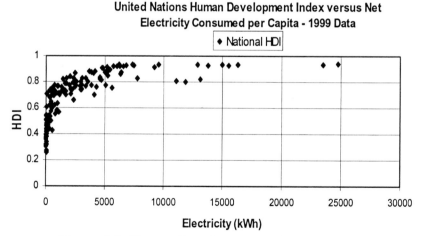

Figure 1-3. Human Development Index (HDI) and Annual Electricity Consumption

The three countries with the largest per capita consumption of electricity in Figure 1-3 are Norway (24773 kilowatt-hours, HDI = 0.939), Iceland (23486 kilowatt-hours, HDI = 0.932), and Canada (16315 kilowatt-hours, HDI = 0.936). The per capita consumption of electricity in the United States (HDI = 0.934) was 12838 kilowatt-hours in 1999.

Figure 1-3 shows that quality of life, as measured by HDI, increases as per capita electricity consumption increases. It also shows that the increase is not linear; the improvement in quality of life begins to level off when per capita electricity consumption rises to about 5000 kilowatt-hours. A similar plot can be prepared for per capita energy consumption (Figure 1-4).

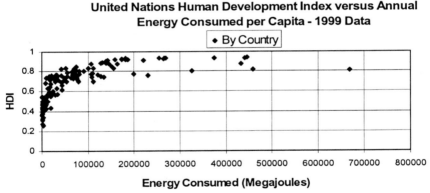

Figure 1-4. HDI and Annual Energy Consumption

Figure 1-4 is a plot of HDI versus per capita energy consumption with all nations with a population of at least one million people. The HDI data are from the 2001 United Nations Human Development Report [UNDP, 2001], and annual per capita energy consumption data are 1999 data reported by the Energy Information Administration of the

United States Department of Energy [EIA Table E.1, 2002]. The figure shows that quality of life increases as per capita energy consumption increases. As in Figure 1-3, the increase is not linear; the improvement in quality of life begins to level off when per capita energy consumption rises to about 200,000 megajoules.

It is interesting to note that some countries have relatively low HDI values, on the order of 80%, despite a relatively large per capita consumption of electricity and energy. These countries include Kuwait (13082 kilowatt-hours, HDI = 0.818), Qatar (11851 kilowatt-hours, HDI = 0.801), and the United Arab Emirates (11039 kilowatt-hours, HDI = 0.809). All of these countries have relatively small populations (less than three million people each in 1999). In addition to their citizenry, the populations in these countries include relatively large, poor service classes. The countries with the largest HDI values, in excess of 90%, are nations with relatively mature economies such as western European nations, Canada, Australia, the United Kingdom, Japan, and the United States. These countries tend to have relatively large middle classes.

Point to Ponder: How can we use quality of life to forecast energy consumption?

The data used to prepare Figures 1-3 and 1-4 can also be used to make a quick forecast of energy demand. Suppose we assume that the world population will stabilize at approximately 8 billion people in the 21st century and that all people will want the quality of life represented by an HDI value of 0.9 (which is approximately the HDI value achieved by Italy, Spain and Israel). In this scenario, the annual per capita energy demand from Figure 1-3 is approximately 200,000 megajoules per person, or a total world energy demand of about 1.6×10^{15} MJ $\approx$ 1500 quads. A quad is a unit of energy that is often used in discussions of global energy because it is comparable in magnitude to global energy values. One quad equals one quadrillion BTU or 10^{15} BTU, or approximately 10^{18} Joules. For comparison, the world population of approximately 6 billion people consumed approximately 387 quads of energy in 1999.

According to our scenario with 8 billion people, worldwide energy demand will almost quadruple by the end of the 21st century when compared to worldwide energy consumption in 1999. Annual per capita energy consumption will have to increase from an average of 68,000 megajoules per person in 1999 to the desired value of 200,000 megajoules per person in 2100.

This calculation illustrates the types of assumptions that must be made to prepare forecasts of energy demand. At the very least, a forecast of demand for energy at the end of the 21st century needs to provide an estimate of the size of the population and the per capita demand for energy at that time. [Fanchi, 2004, Chapter 1]

ENERGY IN TRANSITION

Coal was the first fossil fuel to be used on a large scale. J.U. Nef [1977] described 16th century Britain as the first major economy in the world that relied on coal. Britain relied on wood before it switched to coal. The transition from wood to coal during the period from about 1550 C.E. to 1700 C.E. was made necessary by the excessive consumption of wood that was leading to the eventual deforestation of Britain. Coal was a combustible fuel that could be used as an alternative to wood.

Coal was the fuel of choice during the Industrial Revolution. It was used to boil steam for steam turbines and steam engines. Coal was used in transportation to provide a combustible fuel for steam engines on trains and ships. The introduction of the internal combustion engine made it possible for oil to replace coal as a fuel for transportation. Coal is used today to provide fuel for many coal-fired power plants.

People have used oil for thousands of years [Yergin, 1992, Chapter 1]. Civilizations in the Middle East, such as Egypt and Mesopotamia, collected oil in small amounts from surface seepages as early as 3000-2000 B.C.E. During that period, oil was used in building construction, waterproofing boats and other structures, setting jewels, and mummification. Arabs began using oil to create incendiary weapons as early as 600 C.E. By the 1700's, small volumes of oil were being used in Europe for medicinal purposes and in kerosene lamps. Larger volumes of oil could have been used, but Europe lacked adequate drilling technology.

Pulitzer Prize winner Daniel Yergin [1992, page 20] chose George Bissell of the United States as the person most responsible for creating the modern oil industry. Bissell realized in 1854 that rock oil – as oil was called in the 19th century to differentiate it from vegetable oil and animal fat – could be used as an illuminant. He gathered a group of investors together in the mid-1850's. The group formed the Pennsylvania Rock Oil Company of Connecticut and selected James M. Townsend to be its president.

Bissell and Townsend knew that oil was sometimes produced along with water from water wells. They believed that rock oil could be found below the surface of the Earth by drilling for oil in the same way that water wells were drilled. Townsend commissioned Edwin L. Drake to drill a well in Oil Creek, near Titusville, PA. The location had many oil seepages. The project began in 1857 and encountered many problems. By the time Drake struck oil on Aug. 27, 1859, a letter from Townsend was en route to Drake to inform him that funds were to be cut off [van Dyke, 1997].

Drake's well caused the value of oil to increase dramatically. Oil could be refined for use in lighting and cooking, and became the means for saving whales. The substitution of rock oil for whale oil, which was growing scarce and expensive, reduced the need to hunt whales for fuel to burn in lamps. Within fifteen months of Drake's strike, Pennsylvania was producing 450,000 barrels of oil a year from seventy-five wells. By 1862, three million barrels of oil were being produced and the price of oil dropped to ten cents a barrel [Kraushaar and Ristinen, 1993].

Figure 1-5. Spindletop, Gladys City Boomtown Museum, Beaumont, Texas

The Pennsylvania oil fields provided a relatively small amount of oil to meet demand. In 1882, the invention of the electric light bulb caused a drop in the demand for kerosene. The drop in demand for rock oil was short lived, however. The quickly expanding automobile industry needed oil for fuel and lubrication. New sources of oil were discovered in the early twentieth century. Oil was found in Ohio and Indiana, and later in the San Fernando Valley in California and near Beaumont, Texas. The world's first gusher, a well that produced as much as 75,000 barrels of oil per day, was drilled at Spindletop Hill near Beaumont (Figure 1-5). The well was named Lucas-1 after Anthony F. Lucas, its driller and an immigrant from the Dalmatian coast (now Croatia).

Industrialist John D. Rockefeller began Standard Oil in 1870 and by 1879 the company held a virtual monopoly over oil refining and transportation in the United States. Rockefeller's control of the oil business made him rich and famous. The Sherman Antitrust Act of 1890 was used by the United States government to break Rockefeller's grip on the oil industry. Standard Oil was found guilty of restraining trade and a Federal court ordered the dissolution of Standard Oil in 1909. The ruling was upheld by the United States Supreme Court in 1911 [Yergin, 1992, Ch. 5].

By 1909, the United States produced more oil than all other countries combined, producing half a million barrels per day. Up until 1950, the United States produced more than half of the world's oil supply. Discoveries of large oil deposits in Central and South America, and the Middle East led to decreased United States production. Production in the United States peaked in 1970 and has since been declining. However, oil demand in the United States and elsewhere in the world has continued to grow. Since 1948, the United States has imported more oil than it exports. Today, the United States imports about half of its oil needs.

Until 1973, oil prices were influenced by market demand and the supply of oil that was provided in large part by a group of oil companies called the "Seven Sisters." This group included Exxon, Royal Dutch/Shell, British Petroleum (BP), Texaco, Mobil, Standard Oil of California (which became Chevron), and Gulf Oil. In 1960, Saudi Arabia led the formation of the Organization of Petroleum Exporting Countries, commonly known as OPEC. It was in 1973 that OPEC became a major player in the oil business by raising prices on oil exported by its members. This rise in price became known as the "first oil crisis" as prices for consumers in many countries jumped.

Today, fossil fuels are still the primary fuels for generating electrical power, but society is becoming increasingly concerned about the global dependence on finite resources and the environmental impact of fossil fuel combustion. As a result, society is in the process of changing the global energy mix from an energy portfolio that is heavily dependent on fossil fuels to an energy portfolio that depends on several energy sources. The transition process began in the latter half of the 20th century and is illustrated in Figure 1-6.

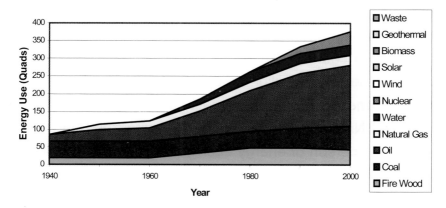

Figure 1-6. Historical Energy Consumption

Figure 1-6 shows total energy consumption for the world in quads from 1940 to 2000. In 1940, the world relied on firewood, coal and oil. Natural gas, energy from water, especially hydropower from dams, and nuclear energy joined firewood, coal and oil as important contributors to the energy mix by the end of the 20th century. Other energy sources – identified as wind, solar, biomass, geothermal and waste in Figure 1-5 – were beginning to make an appearance in the global energy mix at the beginning of the 21st century. They do not appear in the figure because their impact was negligible in the last half of the 20th century. Their possible impact on the future energy mix is discussed later.

The dominance of fossil fuels in the energy mix at the end of the 20th century is illustrated in Figure 1-6. Oil accounted for approximately 22% of world energy consumed in 1940, and for approximately 45% of world energy consumed in 2000.

Point to Ponder: Why should I care about the global distribution of energy?

Suppose a country, like North Korea, has a population of approximately 20 million people. If it wants to provide enough energy to sustain a quality of life corresponding to a United Nations HDI of 0.9, it will require 200,000 mega-joules per person of energy each year. This corresponds to approximately 127 power plants with a 1000 megawatts capacity each. Where will this energy come from? Today, the energy comes primarily from fossil fuels such as oil, gas and coal. In a few countries such as France, it can be provided by nuclear fission energy. If the country does not have significant reserves of fossil fuels or uranium – a material needed for nuclear reactors – it will have to import the materials it needs. In this case the country is a "have not" country that is dependent on countries that have the resources and technology it needs. This creates an opportunity for "have" countries to manipulate "have not" countries. On the other hand, it creates an incentive for "have not" countries to use its human resources to take what is needed. For example, the "have not" country could maintain a large standing army or sponsor acts of violence to influence "have" countries. [Fanchi, 2004, Exercise 1-10]

"DECARBONIZATION"

Energy forecasts rely on projections of historical trends. Table 1-2 is based on historical data presented by the United States Energy Information Administration for the last four decades of the 20th century. The table shows historical energy consumption in units of quads. The row of data labeled "Geothermal, etc." includes net electricity generation from wood, waste, solar, and wind. It is worth noting here that published statistical data are subject to revision, even if the data are historical data that have been published by a credible source. Data revisions may change specific numbers as new information is received and used to update the database, but it is reasonable to expect the data presented in Table 1-2 to show qualitatively correct trends.

Table 1-2. World Primary Energy Production by Source, 1970 – 2000 [EIA Table 11.1, 2002]				
Primary Energy	**Primary Energy Production (quads)**			
	1970	**1980**	**1990**	**2000**
Coal	62.96	72.72	94.29	92.51
Natural Gas	37.09	54.73	75.91	90.83
Crude Oil	97.09	128.12	129.50	145.97
Natural Gas Plant Liquids	3.61	5.10	6.85	9.28
Nuclear Electric Power	0.90	7.58	20.31	25.51
Hydroelectric Power	12.15	18.06	22.55	27.46
Geothermal, etc.	1.59	2.95	3.94	5.36

The data in Table 1-2 are graphically displayed in Figure 1-7. The first four energy sources in Figure 1-7 – coal, natural gas, crude oil, and natural gas plant liquids – are fossil fuels. The data show the dominance of fossil fuels in the energy mix at the end of the 20th century. The data also show that the amount of produced energy from non-fossil fuels is increasing.

The trend in the 20th century has been a "decarbonization" process, that is, a move away from fuels with many carbon atoms to fuels with few or no carbon atoms. H.J. Ausubel [2000, page 18] defines decarbonization as "the progressive reduction in the amount of carbon used to produce a given amount of energy." Figure 1-8 illustrates how the carbon to hydrogen ratio (C:H) declines as the fuel changes from carbon-rich coal to carbon-free hydrogen. The use of hydrogen as a fuel is discussed in more detail later.

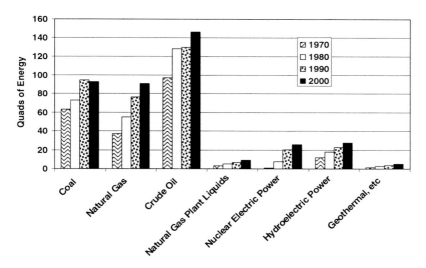

Figure 1-7. Historical Energy Consumption

Figure 1-9 shows the historical pattern and postulated future trends presented by Ausubel [2000]. The figure uses fractional market share M to present the market ratio M/(1-M) of a fuel as a function of time. The historical slopes of the market ratio for coal and oil are assumed to hold true for natural gas and hydrogen during the 21st century. The historical trend suggests that the 21st century will see a gradual transition from a dominance of fossil fuels in the current energy mix to a more balanced distribution of energy options.

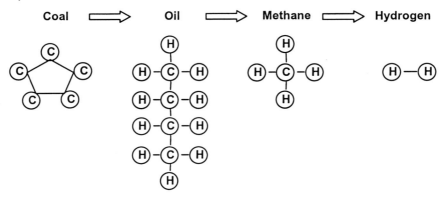

Figure 1-8. Decarbonization

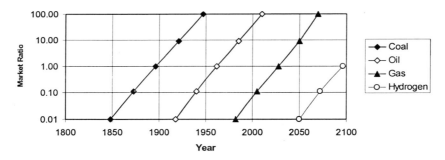

Figure 1-9. Postulated Trends in Global Energy Consumption

Point to Ponder: Why would an oil rich country worry about alternative energy?

The trend toward decarbonization can help explain why a country, like Iran, that is rich in fossil fuels would seek to develop nuclear energy and alternative energy sources. Even if the country is able to export fossil fuels such as oil and gas for centuries to come at its current rate of export, the country can look at forecasts of energy consumption and see that the market is changing. They can use revenue from the sale of fossil fuels to help them transition to new energy sources. The remaining fossil fuels, especially oil, can be used for other applications besides fuel. For example, oil is used in the manufacture of plastics and other refined products such as lubricants.

Chapter 2

ENERGY OPTIONS – FOSSIL ENERGY

The literature contains several publications that present a description of the energy sources that are available or are expected to be available during the 21st century. Energy options known today include fossil fuels, nuclear energy, solar energy, renewable energy and alternative sources. These energy options are briefly described in the chapters that follow.

ORIGIN OF FOSSIL FUELS

Fossil fuels are sources of energy that were formed by the death, decay, and transformation, or diagenesis, of life. The term diagenesis encompasses physical and chemical changes that are associated with lithification and compaction. Sediment can be lithified, or made rock-like, by the movement of minerals into sedimentary pore spaces.

The minerals can form cement that binds grains of sediments together into a rock-like structure that has less
porosity than the original sediment. Porosity is the fraction of
void space between the grains in the material. The volume of
oil or gas in the rock depends on the porosity of the rock
because the fluids are stored in the rock pore space. Two
theories of the origin of fossil fuels are considered here: the
biogenic theory, and the abiogenic theory.

Biogenic Theory

The biogenic theory is the mainstream scientific view of the
origin of fossil fuels. In the biogenic theory, a type of
biochemical precipitation called organic sedimentation forms
coal, oil and gas. When vegetation dies and decays in
aqueous environments such as swamps, it can form a
carbon rich organic material called peat. If peat is buried by
subsequent geological activity, the buried peat is subjected
to increasing temperature and pressure. Peat can eventually
be transformed into coal by the process of diagenesis. A
similar diagenetic process is thought to be the origin of oil
and gas.

Oil and gas are petroleum fluids. A petroleum fluid is
a mixture of hydrocarbon molecules and inorganic impurities,
such as nitrogen, carbon dioxide, and hydrogen sulfide.
Petroleum can exist in solid, liquid or gas form depending on
its composition and the temperature and pressure of its
surroundings. Natural gas is typically methane with lesser
amounts of heavier hydrocarbon molecules like ethane and

propane. The elemental mass content of petroleum fluids ranges from approximately 84% to 87% carbon and 11% to 14% hydrogen. These percentages are comparable to the carbon and hydrogen content of life. This is one piece of evidence that supports the idea that petroleum was formed from biological sources.

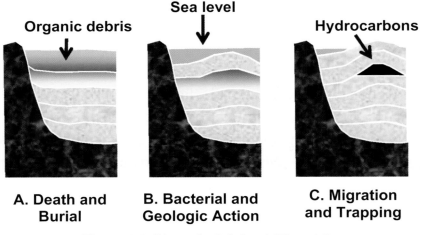

A. Death and Burial   B. Bacterial and Geologic Action   C. Migration and Trapping

Figure 2-1. Biogenic Origin of Oil and Gas

The biochemical process for the formation of petroleum is illustrated in Figure 2-1. It begins with the death of microscopic organisms such as algae and bacteria. The remains of the organisms settle into the sediments at the base of an aqueous environment as organic debris. Lakebeds and seabeds are examples of favorable sedimentary environments. Subsequent sedimentation buries the organic debris. As burial continues, the organic material is subjected to increasing temperature and pressure, and is transformed by bacterial action into oil and gas. Petroleum fluids are usually less dense than water and will migrate upwards until they encounter impermeable barriers and are collected in traps. The accumulation of hydrocarbon fluid in a geologic trap forms a petroleum reservoir.

Abiogenic Theory

In the biogenic theory, the origin of oil and gas begins with the death of organisms that live on or near the surface of the Earth. An alternative hypothesis called the abiogenic theory says that processes deep inside the Earth, in the Earth's mantle, form petroleum. Thomas Gold, an advocate for the abiogenic theory, pointed out that the biogenic theory was adopted in the 1870's. At the time, scientists thought that the Earth was formed from molten rock that was originally part of the Sun. In 1846, Lord Kelvin, a.k.a. William Thomson, estimated the age of the Earth from the rate of cooling of molten rock to be about 100 million years old. A more accurate estimate of the age of the Earth could be made after French physicist Antoine Henri Becquerel discovered radioactivity in 1896. In 1905, Ernest Rutherford proposed using radioactivity to measure the age of the Earth from the concentration of long-lived radioactive materials in rock. The Earth is now believed to be over four billion years old.

Scientists now believe that the Earth was formed by the accumulation and compression of cold nebular material, including simple organic molecules. Gold argues that simple inorganic and organic molecules in the forming Earth were subjected to increasing heat and pressure, and eventually formed more complex molecules. These complex molecules eventually became the simplest forms of life.

Gold's view is consistent with the hypothesis put forward by Russian biochemist A.I. Oparin and British geneticist J.D.S. Haldane that life emerged from inanimate material under the prevailing conditions of the primitive Earth. A classic experiment that supported the Oparin-Haldane hypothesis was performed in 1953 by two American chemists, Stanley Miller and Harold Urey. Their experiment showed that simple inorganic molecules could combine to form some basic molecules of life under conditions that may have been present on the primitive Earth. The interior of the Earth is viewed by proponents of the abiogenic theory as the crucible for forming life. The Miller-Urey experiment can be considered a model of the conditions that existed in the Earth's mantle.

Gold [1999] refutes challenges to the abiogenic theory and presents several pieces of evidence in support of the abiogenic theory, and the possible existence of a biological community deep inside the Earth. Some of Gold's evidence includes the existence of microbial populations that can thrive in extreme heat. These microbes, notably bacteria and archaea, grow at hot, deep ocean vents and can feed on hydrogen, hydrogen sulfide, and methane. Gold considers life forms at deep ocean vents transitional life forms that exist at the interface between two biospheres.

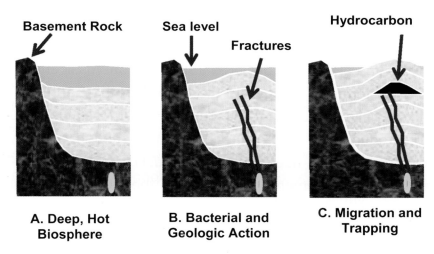

Figure 2-2. Abiogenic Origin of Oil and Gas

One biosphere is the surface biosphere and includes life that lives on the continents and in the seas on the crust of the Earth. Gold postulates that a second biosphere exists in the mantle of the Earth. He calls the second biosphere the deep biosphere. The surface biosphere uses chemical energy extracted from solar energy, while the deep biosphere feeds directly on chemical energy. Oxygen is a requirement in both biospheres. Gold's deep biosphere is the source of life that eventually forms hydrocarbon mixtures (petroleum) in the Earth's mantle. Crustal oil and gas reservoirs are formed by the upward migration of petroleum fluid until the fluid is stopped by impermeable barriers and accumulates in geological traps. The abiogenic theory is illustrated in Figure 2-2.

Point to Ponder: Why does it matter whether the biogenic theory or the abiogenic theory is right?

Two of the arguments driving a transition from fossil energy to other forms of energy are the belief that the Earth contains a finite amount of fossil fuels, and that fossil fuels are not produced by natural processes fast enough to allow fossil fuels to be used as an inexhaustible source of energy. These arguments are based on the assumption that the biogenic theory is correct. If the abiogenic theory is correct, existing estimates of the volume of petroleum and the rate at which it is renewed could be significantly understated. There is still an argument for reducing our dependence on fossil fuels: global warming. Global warming is considered in more detail in Chapter 9.

FOSSIL FUELS

Fossil fuels are the dominant energy source in the modern global economy. They include coal, oil and natural gas. Each of these is discussed below.

Coal

Coal is formed from organic debris by a process known as coalification. When some types of organic materials are heated and compressed over time, they can form water, gas and coal. In some cases, a high-molecular weight, waxy oil is also formed. For example, swamp vegetation may be buried under anaerobic conditions and become peat. Peat is an unconsolidated deposit of partially carbonized vegetable matter in a water-saturated environment such as a bog. If peat is overlain by rock and subjected to increasing temperature and pressure, it can form coal.

Organisms that form coal when subjected to coalification include algae, phytoplankton and zooplankton. Coal can also be formed by the bacterial decay of plants and, to a lesser extent, animals. Organic debris is composed primarily of carbon, hydrogen, and oxygen. It may also contain minor amounts of other elements such as nitrogen and sulfur. The organic origin of coal provides an explanation for the elemental composition of coal, which ranges from pure carbon to a compound of such elements as carbon, hydrogen, oxygen, and sulfur.

Coals are classified by rank. Rank is a measure of the degree of coalification or maturation of the coal. The lowest rank coal is lignite, followed in order by sub-bituminous coal, bituminous coal, anthracite and graphite. Coal rank is correlated to the maturity, or age, of the coal. As a coal matures, the ratio of hydrogen to carbon atoms decreases and the ratio of oxygen to carbon atoms decreases. The highest rank coal, graphite, approaches 100% carbon. Coal becomes darker and denser with increasing rank.

Coals burn better if they are relatively rich in hydrogen; this includes lower rank coals with higher hydrogen to carbon ratios. The percentage of volatile materials in the coal decreases as coal matures. Volatile materials include water, carbon dioxide and methane. Coal gas is gas absorbed in the coal. It is primarily methane with lesser amounts of carbon dioxide. The amount of gas that can be absorbed by the coal depends on the rank. As rank increases, the amount of methane in the coal increases because the molecular structure of higher rank coals has a greater capacity to absorb gas and therefore can contain more gas.

Figure 2-3 shows an idealized representation of the physical structure of a coal seam. A coal seam is the stratum or bed of coal. It is a collection of coal matrix blocks bounded by natural fractures. The fracture network in coalbeds consists of microfractures called "cleats." An inter-connected network of cleats allows coal gas to flow from the coal matrix blocks when the pressure in the fracture declines. This is an important mechanism for coalbed methane production.

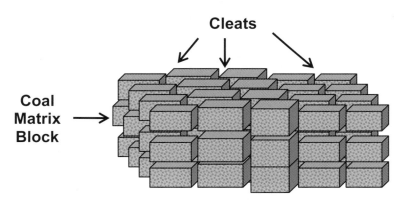

Figure 2-3. Schematic of a Typical Coal Seam

Coal is usually produced by extraction from coal beds. Mining is the most common extraction method. There are several types of mining techniques. Some of the more important coal mining techniques are strip mining, drift mining, deep mining, and longwall mining. Strip mining is also known as surface mining. Coal on the surface of the Earth is extracted by scraping. Drift mines are used to extract coal from coal seams that are exposed by the slope of a mountain. Drift mines typically have a horizontal tunnel entrance into the coal seam. Deep mining extracts coal from beneath the surface of the Earth. In the case of deep mining, coal is extracted by mining the coal seam and leaving the bounding overburden layers and underburden layers undisturbed.

Figure 2-4. Coal Trains in Canada

Coal is transported to consumers by ground trans-portation, especially by trains and, to a lesser extent, ships (Figure 2-4). A relatively inexpensive means of transporting coal is the coal slurry pipeline. Coal slurry is a mixture of water and finely crushed coal. Coal slurry pipelines are not widely used because it is often difficult to obtain rights of way for coal slurry pipelines that extend over long distances, particularly in areas where a coal slurry pipeline would compete with an existing railroad right of way.

Coalbed Methane
Environmental concerns are motivating a change from fossil fuels to an energy supply that is clean. Clean energy refers to energy that has little or no detrimental impact on the environment. Natural gas is a source of relatively clean energy. Oil and gas fields are considered conventional sources of natural gas. Two non-conventional sources of natural gas are coalbed methane, and gas hydrates.

Coalbeds are an abundant source of methane. Coal-bed methane exists as a monomolecular layer (a layer that is one molecule thick) on the internal surface of the coal matrix. Its composition is predominately methane, but can also include other constituents, such as ethane, carbon dioxide, nitrogen and hydrogen. The out-gassing of gas from coal is well known to coal miners as a safety hazard, and occurs when the pressure in the cleat system declines. The meth-ane in the microscopic pore structure of the coalbed may be a safety hazard to miners, but it can also be a source of natural gas. Coal gas is able to diffuse into the natural fracture network when a pressure difference exists between the matrix and the fracture network.

Gas recovery from coalbeds depends on three processes. Gas recovery begins with desorption of gas from the monomolecular layer of gas on the coal surface to the coal micropores. The gas then diffuses through the coal micropores into the cleats. Finally, gas flows through the cleats to the production well. The flow rate depends, in part, on the pressure gradient in the cleats and the density and distribution of cleats. The controlling mechanisms for gas production from coalbeds are the rate of desorption from the coal surface to the coal matrix, the rate of diffusion from the coal matrix to the cleats, and the rate of flow of gas through the cleats.

Petroleum

Petroleum is a mixture of hydrocarbon molecules. Table 2-1 summarizes the composition of petroleum fluids for the most common elements. The actual elemental composition of a petroleum fluid depends on such factors as the composition of its source, reservoir temperature and reservoir pressure.

Hydrocarbon molecules in petroleum fluids are organic molecules. We expect the molecules in petroleum fluids to be relatively non-reactive and stable because they have been present in the fluid mixture for millions of years. If they were reactive or unstable, it is likely that they would have reacted or decomposed at some point in time and their products would be present in the petroleum fluid.

The volume of a petroleum mixture depends on changes in composition as well as changes in temperature and pressure. Fluids that are one phase at reservoir conditions often become two-phase fluids by the time they flow up the wellbore and reach the surface. Natural gas is a petroleum fluid in the gaseous state at surface conditions. Oil is a petroleum fluid in the liquid state at surface conditions. Heavy oils do not contain much gas in solution and have a relatively large molecular weight. By contrast, light oils typically contain a large amount of gas in solution and have a relatively small molecular weight.

Table 2-1 Elemental Composition of Petroleum Fluids	
Element	Composition (% by mass)
Carbon	84% - 87%
Hydrogen	11 % - 14%
Sulfur	0.6% - 8 %
Nitrogen	0.02% - 1.7%
Oxygen	0.08% - 1.8%
Metals	0% - 0.14%

Petroleum fluids are usually found in the pore space of sedimentary rocks. Igneous and metamorphic rocks originated in high pressure and temperature conditions that did not favor the formation or retention of petroleum fluids. Any petroleum fluid that might have occupied the pores of a metamorphic rock is usually cooked away by heat and pressure.

Several key factors must be present to allow the development of a hydrocarbon reservoir:

1. A source for the hydrocarbon must be present. For example, one source of oil and gas is thought to be the decay of single celled aquatic life. Shales formed by the heating and compression of silts and clays are often good source rocks. Oil and gas can form when the remains of an organism are subjected to increasing pressure and temperature.

2. A flow path must exist between the source rock and reservoir rock.

3. Once hydrocarbon fluid has migrated to a suitable reservoir rock, a trapping mechanism becomes important. If the hydrocarbon fluid is not stopped from migrating, buoyancy and other forces will cause it to move towards the surface.

4. Overriding all of these factors is timing. A source rock can provide large volumes of oil or gas to a reservoir, but the trap must exist at the time oil or gas enters the reservoir.

The stages in the life of a reservoir begin when the first discovery well is drilled. Prior to the discovery well, the reservoir is an exploration target. After the discovery well, the reservoir is a resource that may or may not be economic. The production life of the reservoir begins when fluid is withdrawn from the reservoir. Reservoir boundaries are established by seismic surveys and delineation wells. Delineation wells are wells that are originally drilled to define the size of the reservoir, but can also be used for production or injection later in the life of the reservoir. Production can begin immediately after the discovery well is drilled, or years later after several delineation wells have been drilled. The number of wells used to develop the field, the location of the wells, and their flow characteristics are among the many issues that must be addressed by reservoir management.

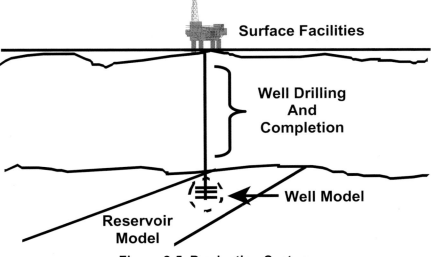

Figure 2-5. Production System

A production system can be thought of as a collection of subsystems illustrated in Figure 2-5. Fluids are taken from the reservoir using wells. Wells must be drilled and completed. The performance of the well depends on the properties of the reservoir rock, the interaction between the rock and fluids in the reservoir, and properties of the fluids in the reservoir. Reservoir fluids include the fluids originally contained in the reservoir as well as fluids that may be introduced as part of the reservoir management process described below. Well performance also depends on the properties of the well itself, such as its cross-section, length, trajectory, and completion. The completion of the well establishes the connection between the well and the reservoir. A completion can be as simple as an open-hole completion where fluids are allowed to drain into the wellbore from consolidated reservoir rock, to completions that require the use of tubing with holes punched through the walls of the tubing using perforating guns to allow fluid to flow between the tubing and the reservoir.

Surface facilities are needed to drill, complete and operate wells. Drilling rigs may be moved from one location to another on trucks, ships, or offshore platforms (Figure 2-6); or drilling rigs may be permanently installed at specified locations. The facilities may be located in desert climates in the Middle East, stormy offshore environments in the North Sea, arctic climates in Alaska and Siberia, and deepwater environments in the Gulf of Mexico and off the coast of West Africa.

Figure 2-6. Offshore Platform in Dry Dock, Galveston, Texas

Produced fluids must be recovered, processed and transported to storage facilities and eventually to the consumer. Processing can begin at the well site where the produced wellstream is separated into oil, water and gas phases. Further processing at refineries separates the hydrocarbon fluid into marketable products, such as gasoline and diesel fuel (Figure 2-7). Transportation of oil and gas may be by a variety of means, including pipelines, tanker trucks, double hulled tankers, and ships capable of carrying liquefied natural gas.

Figure 2-7. A South Texas Refinery

Gas Hydrates

Gas hydrates are chemical complexes that are formed when one type of molecule completely encloses another type of molecule in a lattice. In the case of gas hydrates, hydrogen-bonded water molecules form a cage-like structure in which mobile molecules of gas are absorbed or bound. Although gas hydrates can be found throughout the world, difficulties in cost-effective production have hampered development of the resource. Gas hydrates are generally considered troublesome for oil and gas field operations, but the commercial potential of methane hydrates as a relatively clean energy resource is changing the industry perception of gas hydrates.

Methane hydrates contain a relatively large volume of methane in the hydrate complex. The hydrate complex contains about 85 mole percent water and approximately 15 mole percent guests, where a guest molecule is methane or some other relatively low molecular weight hydrocarbon. Methane hydrates can be found throughout the world. They exist on land in sub-Arctic sediments and on seabeds where the water is near freezing. Difficulties in cost-effective pro-duction of methane hydrates have hampered the production of methane from hydrates.

Tight Gas Sands and Shale Gas

Non-conventional gas resources include coalbed methane, tight gas sands and fractured gas shales. Coalbed methane was discussed above. Tight gas sands and gas shales are characterized by low permeabilities, that is, permeabilities that are a fraction of a millidarcy (less than 10^{-15} m^2). The low permeability associated with non-conventional gas resources makes it more difficult to produce the gas at economical rates.

Economic production of gas from a gas shale or tight gas sand often requires the creation of fractures by a proc-ess known as hydraulic fracturing. In this process, a fluid is injected into the formation at a pressure that exceeds the fracture pressure of the formation. Once fractures have been created in the formation, a proppant such as coarse grain sand or manmade pellets are injected into the fracture to prevent the fracture from closing, or healing, when the injection pressure is removed. The proppant provides a higher permeability flow path for gas to flow to the production well. Non-conventional low permeability gas sands and shales often require more wells per unit area than conventional higher permeability gas reservoirs. The key to managing a non-conventional gas resource is to develop the resource with enough wells to maximize gas recovery without drilling unnecessary wells.

Shale Oil and Tar Sands

Shale oil is contained in porous, low permeability shale. Sand grains that are cemented together by tar or asphalt are called tar sands. Tar and asphalt are highly viscous, plastic or solid hydrocarbons. Extensive shale oil and tar sand deposits are found throughout the Rocky Mountain region of North America, as well as in other parts of the world. Although difficult to produce, the volume of hydrocarbon in tar sands has stimulated efforts to develop production techniques.

The hydrocarbon in shale oil and tar sands can be extracted by mining when oil shales and tar sands are close enough to the surface. Tar pits have been found around the world and have been the source of many fossilized dinosaur bones. In locations where oil shales and tar sands are too deep to mine, it is necessary to increase the mobility of the hydrocarbon.

An increase in permeability or a decrease in viscosity can increase mobility. Increasing the temperature of shale oil, tar or asphalt can significantly reduce viscosity. If there is enough permeability to allow injection, steam or hot water can be used to increase formation temperature and reduce hydrocarbon viscosity. In many cases, however, permeability is too low to allow significant injection of a heated fluid. An alternative to fluid injection is electromagnetic heating. Radio frequency heating has been used in Canada, and electro-magnetic heating techniques are being developed for other parts of the world.

Point to Ponder: How does oil price affect recovery?
We have stated before that many experts believe we are running out of oil. It is becoming increasingly difficult to discover new reservoirs that contain large volumes of oil and gas. Much of the exploration effort is focusing on less hospitable climates, such as arctic conditions in Siberia and deep water, offshore regions near West Africa. Yet we already know where large volumes of oil remain: in the reservoirs we have already discovered and developed. Current development techniques have recovered approxi-mately one third of the oil in known fields. That means roughly two thirds remains in the ground where we found it. [Fanchi, 2004, Exercise 6-9]

The efficiency of oil recovery depends on cost. We can produce much more oil from existing reservoirs if we are willing to pay for it. Most oil producing companies choose to seek and produce less expensive oil so they can compete in the international marketplace. Table 2-2 illustrates the sensitivity of oil producing techniques to the price of oil. The table shows that more sophisticated technologies can be justified as the price of oil increases. It also includes a price estimate for alternative energy sources, such as wind and solar. In some cases there is overlap between one technology and another. For example, steam flooding is an Enhanced Oil Recovery (EOR) process that can compete with conventional oil recovery techniques such as waterflooding, while chemical flooding is an EOR process that can be as expensive as many alternative energy sources.

Table 2-2. Sensitivity of Oil Recovery Technology to Oil Price	
Oil Recovery Technology	Oil Price (US$ per barrel in year 2000 US$)
Conventional	10 – 30
Enhanced Oil Recovery (EOR)	20 – 40
Extra Heavy Oil (e.g. tar sands)	25 – 45
Alternative Energy Sources	40 +

Point to Ponder: How high can oil prices go?
In addition to relating recovery technology to oil price, Table 2-2 contains another important point: the price of oil

cannot rise to an arbitrarily high price without encountering competition from other energy options. For the data given in the table, we see that alternative energy sources become cost competitive when the price of oil rises above US$40 per barrel. If the price of oil stays at US$40 per barrel or higher for an extended period of time, energy consumers will begin to switch to less expensive energy sources. This switch has already begun in some countries. For example, consumers in European countries pay much more for gasoline than consumers in the United States. Countries such as Denmark, Germany and Holland are rapidly developing wind energy as an alternative to fossil fuels. France has opted to rely on nuclear fission energy.

Historically, we have seen oil exporting countries try to maximize their income and minimize competition from alternative energy and expensive oil recovery technologies by supplying just enough oil to keep the price at around US$25 to US$35 per barrel. Oil importing countries can attempt to minimize their dependence on imported oil by developing technologies that reduce the cost of alternative energy. If an oil importing country contains mature oil reservoirs, the development of relatively inexpensive tech-nologies for producing oil remaining in mature reservoirs or the imposition of economic incentives to encourage domestic oil production can be used to reduce the country's dependence on imported oil.

FOSSIL ENERGY AND COMBUSTION

The chemical energy in fossil fuels is released by the process of combustion. The environmental impact of increased emission of combustion byproducts must also be considered. When a carbon based fuel burns, the carbon can react with oxygen to form carbon dioxide or carbon monoxide. If hydrogen is present, as it would be when a hydrocarbon is burned, hydrogen reacts with oxygen to form water.

These are exothermic reactions, that is, reactions that have a net release of energy. An exothermic reaction between two reactants A, B has the form

$$A + B \rightarrow \text{products} + \text{energy}$$

Although the exothermic reaction releases energy, it may actually require energy to initiate the reaction. The energy that must be added to a system to initiate a reaction is called activation energy. An exothermic reaction will form products with a net release of energy. Fossil fuels release a relatively large amount of energy during the combustion process. An endothermic reaction does not release energy. Instead, it requires a net input of energy to form the products of the chemical reaction.

Point to Ponder: Why is fossil fuel combustion considered a problem?

Fossil fuel combustion provided a new source of energy that helped prevent deforestation and now supports a relatively high quality of life in industrialized nations.

Unfortunately, fossil fuel combustion also releases a large amount of carbon dioxide. Carbon dioxide is known as a greenhouse gas. The accumulation of carbon dioxide in the atmosphere tends to trap heat energy in the atmosphere. Many scientists believe that the additional heat is increasing the temperature of the atmosphere and is causing global warming. Global warming is discussed in more detail in Chapter 9.

Chapter 3

ENERGY OPTIONS – NUCLEAR ENERGY

Nuclear energy is presently provided by nuclear fission. Nuclear fission is the process in which a large, unstable nucleus splits into two smaller fragments. It depends on a finite supply of fissionable material.

Nuclear fusion is the combination, or fusing, of two small nuclei into a single larger nucleus. Many scientists expect nuclear energy to be provided by nuclear fusion sometime during the 21st century. Fusion reactions are the source of energy supplied by the Sun. Attempts to harness and commercialize fusion energy have so far been unsuccessful because of the technical difficulties involved in igniting and controlling a fusion reaction. Nevertheless, fusion energy is expected to contribute significantly to the energy mix by the end of the 21st century, even though a prototype commercial-scale nuclear reactor is not expected to exist until 2015 or later [Morrison and Tsipis, 1998]. Both fission and fusion reactions release large amounts of energy, including a significant amount of waste heat that needs to be dissipated and controlled. The decay products of the fission process can be highly radioactive for long periods of time, while the byproducts of the fusion process are relatively safe.

HISTORY OF NUCLEAR ENERGY

Nuclear energy became an important contributor to the global energy mix in the latter half of the 20th century. We obtain nuclear energy from two types of reactions: fission, and fusion. Fission is the splitting of one large nucleus into two smaller nuclei; fusion is the joining of two small nuclei into one large nucleus. In both reactions, significant amounts of energy can be released. Nuclear energy in the past and present energy mix has been provided by nuclear fission reactions. Nuclear fusion is a future technology that should contribute to the 21st century energy mix. We consider here the historical development of nuclear energy.

Discovery of the Nucleus

German physicist Wilhelm Roentgen observed a new kind of radiation called x-rays in 1895. Roentgen's x-rays could pass through the body and create a photograph of the interior anatomy. Frenchman Henri Becquerel discovered radioactivity in 1896 while looking for Roentgen's x-rays in the fluorescence of a uranium salt. French physicist and physician Marie Curie (originally from Warsaw, Poland) and her husband Pierre Curie were the first to report the discovery of a new radioactive element in 1898. They named the element polonium after Marie's homeland. Ernest Rutherford identified the "rays" emitted by radioactive elements and called them alpha, beta, and gamma rays. Today we know that the alpha ray is the helium nucleus, the beta ray is an electron, and the gamma ray is an energetic photon. By 1913, Rutherford and his colleagues at the Cavendish Laboratory in Cambridge had discovered the nucleus by observing alpha particles bombard thin metallic foils. The constituents of the nucleus were identified as the proton and a new, electrically neutral particle, the neutron. James Chadwick discovered the neutron in 1932 while working in Rutherford's laboratory.

The proton and neutron are classified as nucleons, or nuclear constituents. The number of protons in the nucleus is the atomic number of the nucleus. The mass number of the nucleus is the sum of the number of protons and the number of neutrons.

In an electrically neutral atom, the number of negatively charged electrons is equal to the number of positively charged protons. Isotopes are nuclei with the same atomic number, but different numbers of neutrons. Carbon-12 is the isotope of carbon with a mass number of 12 (six protons plus six neutrons). The trio of isotopes that is important in nuclear fusion is the set of isotopes of hydrogen: hydrogen with a proton nucleus, deuterium with a deuteron nucleus (one proton and one neutron) and tritium with a triton nucleus (one proton and two neutrons).

The number of protons in the nucleus of an element is used to classify the element in terms of its positive electric charge. Isotopes are obtained when electrically neutral neutrons are added to or subtracted from the nucleus of an element. Changes in the number of neutrons occur by several mechanisms. Nucleus changing mechanisms include nuclear emission of helium nuclei (alpha particles), electrons (beta particles), or highly energetic photons (gamma rays). An isotope is said to decay radioactively when the number of protons in its nucleus changes. Elements produced by radioactive decay are called decay products.

Several measures of radioactivity exist. The radiation dose unit that measures the amount of radiation energy absorbed per mass of absorbing material is called the Gray (Gy). One Gray equals one Joule of radiation energy absorbed by one kilogram of absorbing material.

A measure of radiation is needed to monitor the biological effects of radiation for different types of radiation. The measure of radiation is the dose equivalent. Dose equivalent is the product of the radiation dose times a qualifying factor. The qualifying factor indicates how much energy is produced in a material as it is traversed by a given radiation. Table 3-1 illustrates several qualifying factors. The alpha particle referred to in Table 3-1 is the helium nucleus.

Table 3-1. Typical Qualifying Factors [after Murray, 2001, pg. 213, Table 16.1]	
x-rays, gamma rays	1
Thermal neutrons (0.025 electron volt)	2
High energy protons	10
Heavy ions, including alpha particles	20

Dose equivalent is expressed in sieverts (Sv) when the dose is expressed in Grays. R.L. Murray [2001, pg 214] wrote that a single, sudden dose of four sieverts can be fatal, while the typical annual exposure to natural and manmade (e.g. medical and dental) radiation is 3.6 millisieverts (a millisievert is one thousandth of a sievert).

Binding energy is the energy needed to separate the nucleus into its constituent nucleons. Energy can be released from a nucleus when the nucleus splits into two comparable fragments. This splitting, or fission, of the nucleus can occur when the mass number is sufficiently large. The separation of a large nucleus into two comparable fragments is an example of spontaneous fission. The large nucleus is called the parent nucleus, and the fission fragments are called daughter nuclei. By contrast, energy is released in the fusion process when two light nuclei with very small mass numbers are combined to form a larger nucleus.

History of Nuclear Power

Physicist Leo Szilard conceived of a neutron chain reaction in 1934. Szilard knew that neutrons could interact with radioactive materials to create more neutrons. If the density of the radioactive material and the number of neutrons were large enough, a chain reaction could occur. Szilard thought of two applications of the neutron chain reaction: a peaceful harnessing of the reaction for the production of consumable energy; and an uncontrolled release of energy (an explosion) for military purposes. Recognizing the potential significance of his concepts, Szilard patented them in an attempt to hinder widespread development of the military capabilities of the neutron chain reaction. This was the first attempt in history to control the proliferation of nuclear technology.

Italian physicist Enrico Fermi (1901-1954) and his colleagues in Rome were the first to bombard radioactive material using low-energy ("slow") neutrons in 1935. The spatial extent of the nucleus is often expressed in terms of a unit called the Fermi, in honor of Enrico Fermi. One Fermi is roughly the diameter of a nucleus and is the range of the nuclear force. The correct interpretation of Fermi's results as a nuclear fission process was provided in 1938 by Lise Meitner and Otto Frisch in Sweden, and Otto Hahn and Fritz Strassmann in Berlin. Hahn and Strassmann observed that neutrons colliding with uranium could cause the uranium to split into smaller elements. This process was called nuclear fission. The fission process produces smaller nuclei from the break up of a larger nucleus, and can release energy and neutrons. If neutrons interact with the nuclei of fissionable material, the neutrons can cause the nuclei to split and produce more neutrons, which can then react with more fissionable material in a chain reaction. Fermi succeeded in operating the first sustained chain reaction on December 2, 1942 in a squash court at the University of Chicago.

The scientific discovery of radioactivity and nuclear fission did not occur in a peaceful society, but in a world threatened by the militaristic ambitions of Adolf Hitler and Nazi Germany. In 1939, Hitler's forces plunged the world into war. Many prominent German scientists fled to the United States and joined an Allied effort to develop the first nuclear weapons. Their effort, known as the Manhattan Project, culminated in the successful development of the atomic bomb. The first atomic bomb was exploded in the desert near Alamogordo, New Mexico in 1945.

By this time, Hitler's Germany was in ruins and there was no need to use the new weapon in Europe. Japan, however, was continuing the Axis fight against the Allied forces in the Pacific and did not seem willing to surrender without first being defeated on its homeland. Such a defeat, requiring an amphibious assault against the Japanese islands, would have cost many lives, both of combatants and Japanese non-combatants. United States President Harry Truman decided to use the new weapon.

The first atomic weapon to be used against an enemy in war was dropped by a United States airplane on the city of Hiroshima, Japan on August 6, 1945. Approximately 130,000 people were killed and 90% of the city was destroyed. When the Japanese government refused to surrender unconditionally, a second atomic bomb was dropped on the Japanese city of Nagasaki on August 9, 1945. The combined shocks of two nuclear attacks and a Soviet invasion of Manchuria on August 9, 1945 prompted the Japanese Emperor to accept the Allies' surrender terms on August 14, 1945 and to formally surrender aboard the USS Missouri (Figure 3-1) on September 2, 1945.

Figure 3-1. The USS Missouri berthed at Pearl Harbor, Hawaii

People throughout the world realized that nuclear weapons had significantly altered the potential consequences of an unlimited nuclear war. By the early 1950's, the Soviet Union had acquired the technology for building nuclear weapons and the nuclear arms race was on. As of 1972, the United States maintained a slight edge over the Soviet Union in strategic nuclear yield and almost four times the number of deliverable warheads. Yield is a measure of the explosive energy of a nuclear weapon. It is often expressed in megatons, where one megaton of explosive is equivalent to one million tons of TNT.

During the decade from 1972 to 1982, the size of the United States nuclear arsenal did not change significantly, but the quality of the weapons, particularly the delivery systems such as missiles, increased dramatically. By 1982, the Soviet Union had significantly closed the gap in the number of deliverable warheads and had surged ahead of the United States in nuclear yield. The Soviet Union and the United States had achieved a rough nuclear parity. The objective of this parity was a concept called "deterrence."

Deterrence is the concept that neither side will risk engaging in a nuclear war because both sides would suffer unacceptably large losses in life and property. This is the concept that underlies the doctrine of Mutual Assured Destruction: the societies of all participants in a nuclear war would be destroyed. The "star wars" concept advanced by United States President Ronald Reagan in the 1980's is a missile defense system that provided an alternative strategy to Mutual Assured Destruction, but threatened the global balance of nuclear power with the Soviet Union.

The Soviet Union was unable to match the development of "star wars" military technology and still compete economically with the United States. The Soviet Union dissolved into separate states in the late 1980's. Some of the states of the former Soviet Union, notably Russia and Ukraine, retain nuclear technology. Other nations around the world have developed nuclear technology for peaceful purposes, and possibly for military purposes as well.

Point to Ponder: Can we put the nuclear genie back in the bottle?

Some historians argue that nuclear weapons were not needed to end World War II and that the world would be better off if nuclear energy had never been developed. Today, a nation with a nuclear weapons arsenal must be concerned about the security of every weapon in its arsenal. Even one nuclear weapon can be a weapon of mass destruction. In addition, many nations are seeking to acquire nuclear technology, which increases the risk of nuclear weapons proliferation. Now that the world knows about nuclear energy, society must learn to govern its use. One forecast discussed in Chapter 10 is based on the premise that nuclear technology can be safely controlled and used as a long-term source of energy.

NUCLEAR REACTORS

Nuclear reactors are designed for several purposes. The primary commercial purpose is to generate electric power. Nuclear reactors also provide power for ships such as submarines and aircraft carriers, and they serve as facilities for training and research. Our focus here is on the use of nuclear reactors to generate electricity. We will then discuss global dependence on nuclear energy.

Nuclear Fission Reactors

The neutrons produced in fission reactions typically have energies ranging from 0.1 million electron volts (1.6×10^{-14} Joules) to 1 million electron volts (1.6×10^{-13} Joules). Neutrons with energies this high are called fast neutrons. Fast neutrons can lose kinetic energy in collisions with other materials in the reactor. Less energetic neutrons with kinetic energies on the order of the thermal energy of the reactor are called slow neutrons or thermal neutrons. Some nuclides tend to undergo a fission reaction after they capture a slow neutron. These nuclides are called fissile materials and are valuable fuels for nuclear fission reactors. Fission is induced in fissile materials by slow neutrons. Fission reactors that depend on the nuclear capture of fast neutrons exist, but most commercial reactors are currently designed to use slow neutrons.

Moderating materials like light water (light water is ordinary water or H_2O), heavy water (deuterium oxide or D_2O), graphite and beryllium control the number of neutrons in the reactor. They are called moderating materials because they moderate, or control, the nuclear reaction. Moderating materials slow down fast neutrons by acquiring kinetic energy in collisions.

Coolant materials transport the heat of fission away from the reactor core. Coolants include light water, carbon dioxide, helium, and liquid sodium. In some cases, moderating materials can function as coolants.

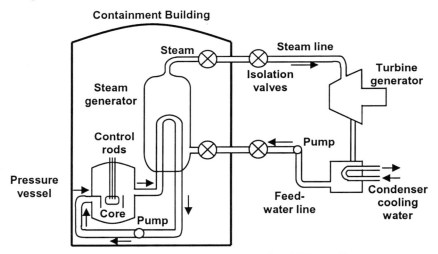

Figure 3-2. Schematic of a Pressurized Water Reactor

A pressurized water reactor is sketched in Figure 3-2. The pressurized water reactor heats coolant water to a high temperature and then sends it to a heat exchanger to produce steam. A boiling water reactor provides steam directly. The steam from each of the reactors turns a turbine to generate electricity. A nuclear reactor is housed in a containment building (Figure 3-3). The containment building, which is typically a dome, provides protection from internal leaks as well as external dangers, such as an airplane crash.

Figure 3-3. Containment Building at San Onofre Nuclear Generating Station, California

Nuclear fission reactor design and operation depends on the ratio of neutrons in succeeding generations of decay products. The ratio of the number of neutrons in one generation to the number of neutrons in the preceding generation is called the neutron multiplication factor. The neutron multiplication factor is a positive quantity that depends on several factors [Lilley, 2001, Chapter 10].

If the multiplication factor for a reactor is less than one, the number of fission reactions will diminish from one generation to the next and the chain reaction will die out. The reactor level is considered subcritical. A critical mass of fissile material is present in the reaction chamber when there are enough neutrons being produced by nuclear reactions to balance the loss of neutrons. When the reaction is at the critical level, the multiplication factor for a reactor is equal to one and the chain reaction can be sustained. If the multiplication factor is greater than one, the number of neutrons produced by nuclear reactions exceeds the loss of neutrons and the number of fission reactions will increase from one generation to the next. The reactor level is considered supercritical. The chain reaction in a supercritical reactor will accelerate and, if it proceeds too fast, can have explosive consequences. The nuclear fission bomb is a supercritical reaction.

Availability of Nuclear Fuel

The most abundant fuel for nuclear fission is uranium. Uranium exists in the crust of the Earth as the mineral uraninite. Uraninite is commonly called pitchblende and is a uranium oxide (U_3O_8). It is found in veins in granites and other igneous rocks. It is possible to find uranium in sedimentary rocks. In this case, scientists believe that uraninite was precipitated in sedimentary rocks after being transported from igneous rocks by the flow of water containing dissolved uraninite.

Uranium is obtained by mining the mineral uraninite. Mining methods include underground mining, open pit mining, and in situ leaching. Leaching is a process of selectively extracting a metal by a chemical reaction that creates a water-soluble molecule that can be transported to a recovery site. The isotope of uranium that undergoes spontaneous fission (uranium-235) is approximately 0.7% of naturally occurring uranium ore. Uranium must be separated from mined ore and then enriched for use in nuclear fission reactors. The enrichment process is designed to purify uranium-235.

Other fuels that can be used in the fission process include the fission products plutonium-239 and thorium-232. Specialized reactors called breeder reactors are designed to operate with fuels other than uranium. A breeder reactor is a nuclear fission reactor that produces more fissile material than it consumes.

The amount of uranium that can be recovered from the Earth is called uranium reserves. Estimates of uranium reserves have been made and have many of the same uncertainties associated with estimates of fossil fuel reserves. Factors that are not well known include the distribution and extent of uranium deposits and the price people are willing to pay to recover the resource. Uranium is considered a nonrenewable resource because it exists as a finite volume within the Earth. Table 3-2 presents uranium reserves estimates provided by the United States Energy Information Administration for the most common mining methods. One of the appealing features of nuclear fusion is the relative abundance of hydrogen and its isotopes compared to fissile materials such as uranium. The table shows that uranium oxide reserves are greater if you are willing to pay for it. The amount of uranium oxide that is left to be produced, like the amount of remaining oil to be produced, depends on price.

Table 3-2. Uranium Reserves by Mining Method, 2001 Estimate Source: Table 3, EIA website, 2002		
Mining Method	US$30 per pound	US$50 per pound
	Uranium Oxide (U_3O_8) million pounds (million kg)	Uranium Oxide (U_3O_8) million pounds (million kg)
Underground	138 (62.6)	464 (210.5)
Open Pit	29 (13.2)	257 (116.6)
In Situ Leaching	101 (45.8)	174 (78.9)
TOTAL	268 (121.6)	895 (406)

Table 3-2 presents a uranium reserve estimate of 268 million pounds at a price of US$30 per pound. The mass fraction of uranium in U_3O_8 is about 0.848 and there is about 0.7% uranium-235 in the uranium oxide. We therefore estimate a uranium-235 reserve of approximately 1.6 million pounds. If we observe that 142 kilograms of uranium-235 can fuel a reactor that outputs 100 megawatts in a year, there is enough uranium-235 reserve at the lower price to operate 500 equivalent reactors for 10 years.

Nuclear Fusion Reactors

The idea behind nuclear fusion is quite simple: fuse two molecules together and release large amounts of energy in the process. Examples of fusion reactions include collisions between protons, deuterons (deuterium nuclei), and tritons (tritium nuclei). Protons are readily available as hydrogen nuclei. Deuterium is also readily available. Ordinary water contains approximately 0.015 mole % deuterium [Murray, 2001, page 77], thus one atom of deuterium is present in ordinary water for every 6700 atoms of hydrogen.

The fusion reaction can occur only when the atoms of the reactants are heated to a temperature high enough to strip away all of the atomic electrons and allow the bare nuclei to fuse. The state of matter containing bare nuclei and free electrons at high temperatures is called plasma. Plasma is an ionized gas. The temperatures needed to create plasma and allow nuclear fusion are too high to be contained by conventional building materials. Two methods of confining plasma for nuclear fusion are being considered.

Magnetic confinement is the first confinement method and relies on magnetic fields to confine the plasma. The magnetic confinement reactor is called a tokamak reactor. Tokamak reactors are toroidal (donut shaped) magnetic bottles that contain the plasma that is to be used in the fusion reaction. Two magnetic fields confine the plasma in a tokomak: one is provided by cylindrical magnets that create a toroidal magnetic field; and the other is a poloidal magnetic field that is created by the plasma current. Combining these two fields creates a helical field that confines the plasma. Existing tokamak reactors inject deuterium and tritium into the vacuum core of the reactor at very high energies. Inside the reactor, the deuterium and tritium isotopes lose their electrons in the high energy environment and become plasmas. The plasmas are confined by strong magnetic fields until fusion occurs.

The second confinement method is inertial confinement. Inertial confinement uses pulsed energy sources such as lasers to concentrate energy onto a small pellet of fusible material, such as a frozen mixture of deuterium and tritium. The pulse compresses and heats the pellet to ignition temperatures.

Point to Ponder: Why is nuclear fusion so desirable?
Nuclear fusion is considered an environmentally clean source of energy. Unlike nuclear fission, nuclear fusion does not generate waste products that are lethal for thousands of years. The material needed for nuclear fusion is much more plentiful than the known supply of fissionable materials, such as uranium. [Fanchi, 2004, Exercises 11-5 and 11-9].

Point to Ponder: When will nuclear fusion be commercially available?

Although many fusion reactions are possible, a commercial fusion reactor has not yet been constructed. Nuclear fusion reactors are still in the research stage and are not expected to provide commercial power for at least a generation. Researchers are still learning how to ignite and sustain nuclear fusion reactions. Nuclear fusion must be controlled before it can be commercialized.

GLOBAL DEPENDENCE ON NUCLEAR POWER

The first commercial nuclear power plant was built on the Ohio River at Shippingport, Pennsylvania, a city about 25 miles from Pittsburgh. It began operation in 1957 and generated 60 megawatts of electric power [Murray, 2001, page 202]. Today, nuclear power plants generate a significant percentage of electricity in some countries. Table 3-3 lists the top ten producers of electric energy from nuclear energy. The table also shows their percentage of the world's total electric energy production from nuclear energy for the year 2000. The source of these statistics is the website of the Energy Information Administration of the United States Department of Energy. These statistics should be viewed as approximate. They are presented here to indicate the order of magnitude of electric power generated by power plants that use nuclear energy as their primary energy source. According to these statistics, we see that the total amount of electric energy generated from nuclear energy in the world was 2,434 billion kilowatt-hour in 2000. Total electric energy consumed in the world that same year was 13,719 billion kilowatt-hour. Electric energy generated from nuclear energy provided approximately 17.7% of the electricity consumed in the world in 2000.

Table 3-3. Top Ten Producers of Electric Energy from Nuclear Energy in 2000 Source: Table 2.7, EIA website, 2002		
Country	Electric Energy from Nuclear Energy (billion kWh)	% Total Electric Energy Produced from Nuclear Energy (World = 2,434 billion kWh)
United States	753.9	30.97%
France	394.4	16.20%
Japan	293.8	12.07%
Germany	161.2	6.62%
Russia	122.5	5.03%
South Korea	103.5	4.25%
United Kingdom	81.7	3.36%
Ukraine	71.1	2.92%
Canada	68.7	2.82%
Spain	58.9	2.42%

Some countries are highly dependent on nuclear energy. Table 3-4 shows the percentage of electric energy generated from nuclear energy compared to total electricity generation for the top ten producers of electric energy from nuclear energy for the year 2000. According to the EIA statistics, most of the electricity generated in France was generated from nuclear energy.

Table 3-4. Dependence of Nations on Nuclear Energy in 2000 Source: Table 6.3, EIA website, 2002			
Country	Electric Energy from Nuclear Energy (billion kWh)	Total Electricity Generation (billion kWh)	Nuclear Share (% National Total)
United States	753.9	3799.9	19.8%
France	394.4	513.9	76.7%
Japan	293.8	1014.7	29.0%
Germany	161.2	537.3	30.0%
Russia	122.5	835.6	14.7%
South Korea	103.5	273.2	37.9%
United Kingdom	81.7	355.8	23.0%
Ukraine	71.1	163.6	43.5%

Table 3-4. Dependence of Nations on Nuclear Energy in 2000 Source: Table 6.3, EIA website, 2002			
Country	Electric Energy from Nuclear Energy (billion kWh)	Total Electricity Generation (billion kWh)	Nuclear Share (% National Total)
Canada	68.7	576.2	11.9%
Spain	58.9	211.6	27.8%

Point to Ponder: What has hindered the global adoption of nuclear energy?

Nuclear energy is a long-term source of abundant energy that has environmental advantages. The routine operation of a nuclear power plant does not produce gaseous pollutants or greenhouse gases like carbon dioxide and methane. Despite its apparent strengths, the growth of the nuclear industry in many countries has been stalled by the public perception of nuclear energy as a dangerous and environmentally undesirable source of energy. This perception began with the use of nuclear energy as a weapon (Figure 3-4) and has been reinforced by widely publicized accidents at two nuclear power plants; Three Mile Island, Pennsylvania and Chernobyl, Ukraine. There are significant environmental and safety issues associated with nuclear energy, particularly nuclear fission. These issues are discussed in more detail in Chapter 9.

**Figure 3-4. "Mushroom Cloud" Associated with the
Detonation of a Nuclear Weapon**

Chapter 4

ENERGY OPTIONS – SOLAR ENERGY

Renewable energy is energy obtained from sources at a rate that is less than or equal to the rate at which the source is replenished. In the case of solar energy, we can only use the energy that is provided by the Sun. Since the remaining lifetime of the Sun is measured in millions of years, many people consider solar energy an inexhaustible supply of energy. In fact, solar energy from the Sun is finite, but should be available for use by many generations of people. Solar energy is therefore considered renewable. Energy sources that are associated with solar energy, such as wind and biomass, are also considered renewable.

Solar radiation may be converted to other forms of energy by several conversion processes. Thermal conversion relies on the absorption of solar energy to heat a cool surface. Biological conversion of solar energy relies on photosynthesis. Photovoltaic conversion generates electrical power by the generation of an electrical current. Wind power and ocean energy conversion rely on atmospheric pressure gradients and oceanic temperature gradients to generate electrical power.

Solar energy is available in three forms: passive, active, and electric. Passive and active solar energy are generally used for space conditioning, such as heating and cooling, while solar electric energy is used to generate electrical power. Each of these forms is discussed below.

SOURCE OF SOLAR ENERGY

Solar energy is energy emitted by a star. Figure 4-1 shows the anatomy of a star. Energy emitted by a star is generated by nuclear fusion. The fusion process occurs in the core, or center, of the star. Energy released by the fusion process propagates away from the core by radiating from one atom to another in the radiation zone of the star. As the energy moves away from the core and passes through the radiation zone, it reaches the part of the star where energy continues its journey towards the surface of the star as heat associated with thermal gradients. This part of the star is called the convection zone. The surface of the star, called the photosphere, emits light in the visible part of the electromagnetic spectrum. The star is engulfed in a stellar atmosphere called the chromosphere. The chromosphere is a layer of hot gases surrounding the photosphere.

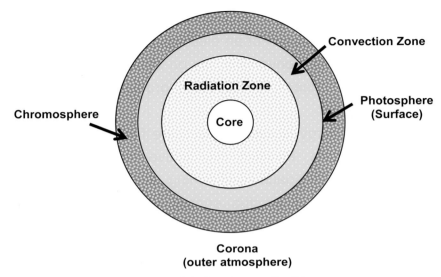

Figure 4-1. Anatomy of a Star

The amount of sunlight that reaches the Earth depends on the motion of the Earth around the Sun. Figure 4-2 illustrates the orbit of the Earth around the Sun. Most planetary orbits lie in the ecliptic plane. The ecliptic plane is the plane of the orbit that intersects the Sun. It is shown in Figure 4-2. The line of intersection between the orbital plane and the ecliptic plane is the line of nodes.

The luminosity of a star is the total energy radiated per second by the star. The amount of radiation from the Sun that reaches the Earth's atmosphere is called the solar constant. The solar constant varies with time because the Earth follows an elliptical orbit around the Sun and the axis of rotation of the Earth is inclined relative to the plane of the Earth's orbit. The distance between points on the surface of the Earth and the Sun varies throughout the year.

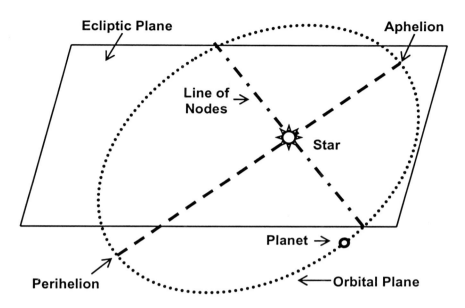

Figure 4-2. Planetary Orbit and the Ecliptic Plane

The amount of solar radiation that reaches the surface of the Earth depends on the factors illustrated in Figure 4-3. The flux of solar radiation incident on a surface placed at the edge of the Earth's atmosphere depends on the time of day and year, and the geographical location of the surface.

Some incident solar radiation is reflected by the Earth's atmosphere. The fraction of solar radiation that is reflected back into space by the Earth-atmosphere system is called the albedo. Approximately thirty-five percent of the light from the Sun does not reach the surface of the Earth. This is due to clouds (20%), atmospheric particles (10%), and reflection by the Earth's surface (5%). The solar flux that enters the atmosphere is reduced by the albedo. Once in the atmosphere, solar radiation can be absorbed in the atmosphere or scattered away from the Earth's surface by atmospheric particulates such as air, water vapor, dust particles, and aerosols. Some of the light that is scattered by the atmosphere eventually reaches the surface of the Earth as diffused light. Solar radiation that reaches the Earth's surface from the disk of the Sun is called direct solar radiation if it has experienced negligible change in its original direction of propagation.

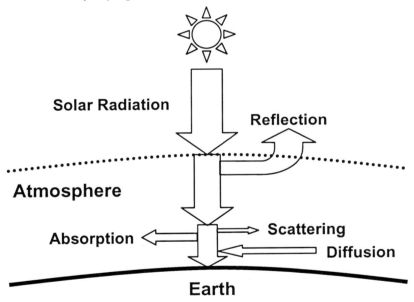

Figure 4-3. Solar Radiation and the Earth-Atmosphere System

Point to Ponder: Can we collect light from outer space?
We have seen that the atmosphere reduces the amount of

sunlight reaching the surface of the Earth. It is conceivable to collect sunlight above the surface of the Earth and transmit it as an intense beam through the atmosphere. The goal would be to reduce the effective albedo of the atmosphere. Obviously, this type of space-age technology will be expensive. There would also be environmental and safety concerns for anything crossing the path of the beam.

PASSIVE SOLAR ENERGY

Passive solar energy technology integrates building design with environmental factors that enable the capture or exclusion of solar energy. Mechanical devices are not used in passive solar energy applications. We illustrate passive solar energy technology by considering two simple but important examples: the roof overhang and thermal insulation.

Roof Overhang

Sunlight that strikes the surface of an object and causes an increase in temperature of the object is an example of direct solar heat. Direct solar heating can cause an increase in temperature of the interior of buildings with windows. The windows that allow in the most sunlight are facing south in the northern hemisphere and facing north in the southern hemisphere. Figure 4-4 illustrates two seasonal cases. The figure shows that the maximum height of the Sun in the sky varies from season to season because of the angle of inclination of the Earth's axis of rotation relative to the ecliptic plane. The Earth's axis of rotation is tilted 23.5° from a line that is perpendicular to the ecliptic plane.

One way to control direct solar heating of a building with windows is to build a roof overhang. The roof overhang is used to control the amount of sunlight entering the windows. Figure 4-4 illustrates a roof overhang. The length of the roof overhang needs to account for the angle α_S of the Sun during the summer and the angle α_W of the Sun during the winter.

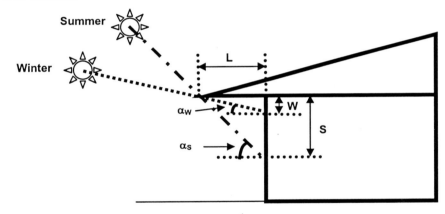

Figure 4-4. Roof Overhang

Passive solar cooling is achieved when the roof overhang casts a shadow over the windows facing the Sun. In this case, the roof overhang is designed to exclude sunlight, and its associated energy, from the interior of the building. Alternatively, the windows may be tinted with a material that reduces the amount of sunlight entering the building. Another way to achieve passive solar cooling is to combine shading with natural ventilation.

Passive solar heating is the capture and conversion of solar energy into thermal energy. The technology for passive solar heating can be as simple as using an outdoor clothes-line to dry laundry or designing a building to capture sunlight during the winter. In the latter case, the building should be oriented to collect sunlight during cooler periods. Sunlight may enter the building through properly positioned windows that are not shaded by a roof overhang, or through skylights. The sunlight can heat the interior of the building and it can provide natural light. The use of sunlight for lighting pur-poses is called daylighting. An open floor plan in the building interior maximizes the effect of daylighting and can substan-tially reduce lighting costs.

Thermal Conductivity and Insulation

Solar energy may be excluded from the interior of a structure by building walls that have good thermal insulation. The quality of thermal insulation for a wall with the geometry shown in Figure 4-5 can be expressed in terms of thermal conductivity and thermal resistance.

The rate of heat flow through the insulated wall shown in Figure 4-5 depends on wall thickness h_{wall}, the cross-sectional area A transverse to the direction of heat flow, and the temperature difference between the exposed and inside faces of the wall. The rate of heat flow through the insulated wall depends on a property of the wall called the thermal conductivity. Thermal conductivity is a measure of heat flow through a material. Metals have relatively high thermal conductivities.

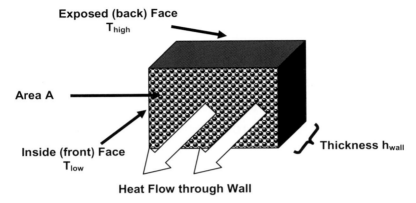

Exposed (back) Face T_{high}

Area A

Inside (front) Face T_{low}

Thickness h_{wall}

Heat Flow through Wall

Figure 4-5. Thermal Conductivity of an Insulated Wall

We have considered thermal insulation as a passive solar technology example. Thermal insulation is also an energy conservation technology. Thermal insulation in walls can keep heat out of a structure during the summer and keep heat in during the winter. Consequently, thermal insulation can reduce the demand for energy to cool a space during the summer and heat a space during the winter. This reduces the demand for energy and makes it possible to conserve, or delay, the use of available energy.

ACTIVE SOLAR ENERGY

Active solar energy refers to the design and construction of systems that collect and convert solar energy into other forms of energy such as heat and electrical energy. Active solar energy technologies are typically mechanical systems that are used to collect and concentrate solar energy. We will discuss solar heat collectors and a solar power plant as illustrations of active solar energy technology.

Solar Heat Collectors

Solar heat collectors capture sunlight and transform radiant energy into heat energy. Figure 4-6 is a diagram of a solar heat collector. Sunlight enters the collector through a window made of a material like glass or plastic. The window is designed to take advantage of the observation that sunlight is electromagnetic radiation with a distribution of frequencies. The window in a solar heat collector is transparent to incident solar radiation and opaque to infrared radiation.

The heat absorber plate in the solar heat collector is a dark surface, such as a blackened copper surface, that can be heated by the absorption of solar energy. The surface of the heat absorber plate emits infrared radiation as it heats up. Sunlight enters through the window, is absorbed by the heat absorber plate, and is reradiated in the form of infrared radiation. Greenhouses work on the same principle: the walls of a greenhouse allow sunlight to enter and then trap reradiated infrared radiation. The window of the solar heat collector is not transparent to infrared radiation so the infrared radiation is trapped in the collector.

The solar heat collector must have a means of transferring collected energy to useful energy. A heat transfer fluid such as water is circulated through the solar heat collector in Figure 4-6 and carries heat away from the

solar heat collector for use elsewhere. Figure 4-7 illustrates
a solar heating system for residential or commercial use.

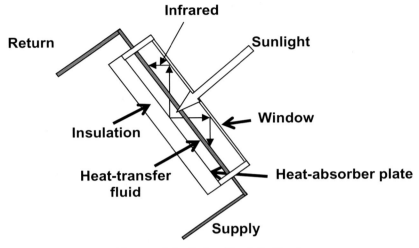

Figure 4-6. Solar Heat Collector

The solar heating system sketched in Figure 4-7 uses
solar energy to heat a liquid coolant such as water or anti-
freeze. The heat exchanger uses heat from the liquid coolant
in the primary circulation system to heat water in the secondary circulation system. The control valve in the lower right
of the figure allows water to be added to the secondary
circulation system. An auxiliary heater in the upper right of
the figure is included in the system to supplement the supply
of heat from the solar collector. It is a reminder that solar
energy collection is not a continuous process. A supplemental energy supply or a solar energy storage system must
be included in the design of the heating system to assure
continuous availability of heat from the solar heating system.

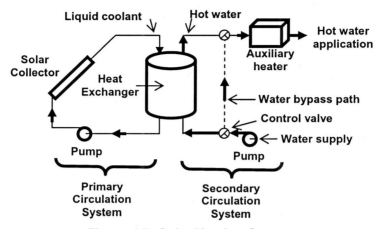

Figure 4-7. Solar Heating System

Energy Conversion Efficiency

The temperature of a solar heat collector does not increase indefinitely because the window and walls of the solar heat collector cannot prevent energy from escaping by convection and radiation. The collector will emit thermal radiation when its temperature is greater than ambient temperature. The energy balance must include energy output as well as energy loss. The energy conversion efficiency of the solar heat collector is the ratio of the energy output by the collector to the energy input to the collector. The energy conversion efficiency depends on the increase in temperature relative to ambient temperature, the intensity of solar radiation, and the quality of thermal insulation. The loss of energy by convection and radiation causes a decrease in energy conversion efficiency.

SOLAR POWER PLANTS

Society is beginning to experiment with solar power plants and a few are in commercial operation. Solar power plants are designed to provide electrical power on the same scale as plants that rely on nuclear or fossil fuel. They use reflective materials like mirrors to concentrate solar energy. The solar power tower and the Solar Electric Generating Station in Southern California are examples of solar power plants. They are described below.

Solar Power Tower

Figure 4-8 is a sketch of a solar power tower with a heliostat field. The heliostat field is a field of large, Sun-tracking mirrors called heliostats arranged in rings around a central receiver tower. The heliostats concentrate sunlight on a receiver at the top of the tower. The solar energy heats a fluid inside the receiver.

Figure 4-9 is a sketch of the geometry of the Sun-tracking mirrors relative to the central receiving station. The heliostats must be able to rotate to optimize the collection of light at the central receiving station. Computers control heliostat orientation. As a ring of heliostats gets farther away from the tower, the separation between the ring and adjacent, concentric rings must increase to avoid shading one ring of mirrors by an adjacent ring.

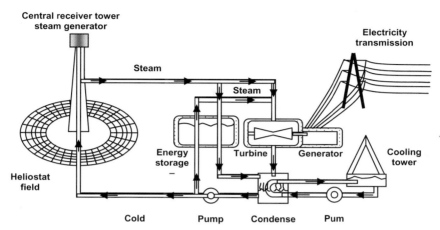

Figure 4-8. Solar Power Tower

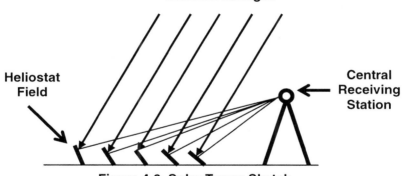

Figure 4-9. Solar Tower Sketch

The first solar power plant based on the solar power tower concept was built in the Mojave Desert near Barstow, California in the 1980's. The solar-thermal power plant at Barstow used 1900 heliostats to reflect sunlight to the receiver at the top of a 300-foot tall tower. The sunlight generates heat to create steam. The steam is used to drive a turbine or it can be stored for later use. The first solar power tower, Solar One, demonstrated the feasibility of collecting solar energy and converting it to electrical energy. Solar One was a 10 megawatt power plant. The heat transfer fluid in Solar One was steam. The Solar One installation was modified to use molten nitrate salt as the heat transfer fluid. The modified installation, called Solar Two, was able to improve heat transfer efficiency and thermal storage for the 10 megawatt demonstration project. The hot salt could be retrieved when needed to boil water into steam to drive a generator turbine.

Solar Electric Generating Systems

A Solar Electric Generating System (SEGS) consists of a large field of solar heat collectors and a conventional power plant. The SEGS plant in Southern California uses rows of parabolic trough solar heat collectors. The collectors are Sun-tracking reflector panels, or mirrors (Figure 4-10). The sunlight reflected by the panels is concentrated on tubes carrying heat transfer fluid. The fluid is heated and pumped through a series of heat exchangers to produce superheated steam. The steam turns a turbine in a generator to produce electricity.

For extended periods of poor weather, solar power plants must use auxiliary fuels in place of sunlight. A prototype SEGS plant used natural gas as an auxiliary fuel. B.Y. Goswami, et al. [2000, Section 8.7] reported that, on average, 75% of the energy used by the plant was provided by sunlight, and the remaining 25% was provided by natural gas. They further reported that solar collection efficiencies ranged from 40% to 50%, electrical conversion efficiency was on the order of 40%, and the overall efficiency for solar to electrical conversion was approximately 15%.

Figure 4-10. Solar Mirrors near Barstow, California

The overall efficiency of a solar electric generating system is the product of optical efficiency, thermal conversion efficiency, and thermodynamic efficiency. Optical efficiency is a measure of how much sunlight is reflected into the system. Thermal conversion efficiency is a measure of how much sunlight entering the system is converted to heat in the system. The thermodynamic efficiency is a measure of how much heat in the system is converted to the generation of electricity.

SEGS plants are designed to supply electrical power to local utilities during peak demand periods. In Southern California, a peak demand period would be hot summer afternoons when the demand for air conditioning is high. This is a good match for a SEGS plant because solar intensity is high. Peak demand periods also correspond to periods of high pollution. One benefit of a SEGS plant is its ability to provide electrical power without emitting fossil fuel pollutants such as nitrous oxide (a component of smog) and carbon dioxide (a greenhouse gas).

Point to Ponder: Why aren't solar power plants more popular in sunny, deserted areas?

One of the major advantages of solar power plants is that the energy is free. There are, however, several issues to consider that impact cost and social acceptability. The free energy from the Sun has to be collected and transformed into commercially useful energy. Solar power plants to date cover the area of several football fields and produce approximately 1% of the power associated with a fossil fuel fired power plant. This means that solar power plants will cover relatively large areas and may be considered eyesores by some people. In addition, the technology of maintaining the collectors, and collecting, transforming, and transmitting solar energy is still relatively expensive. [Fanchi, 2004, Exercises 7-11 and 7-12]

SOLAR ELECTRIC TECHNOLOGY

Solar electric technologies are designed to convert light from the Sun directly into electrical energy. Some of the most important solar electric processes are the photoelectric effect and photovoltaics. We shall briefly discuss these topics below.

Photoelectric Effect

Electrons can move in metals and interact with light. Light is electromagnetic radiation. It was known in the late 1800's that electrons could be ejected from a metal exposed to light, but the effect depended on the frequency of the light. The effect was called the photoelectric effect. Albert Einstein used the concept of a quantum of energy to explain the photoelectric effect in 1905, the same year he published his special theory of relativity. As a historical note, Einstein received the Nobel Prize for his work on the photoelectric effect, not for his theory of relativity.

An electron cannot be ejected from a metal unless it has enough energy to overcome the work function of the metal. The work function is the smallest energy needed to extract an electron from the metal. The work function depends on the type of material and the condition of its surface. Einstein postulated that a collision between an electron and a particle of light called a photon could transfer enough energy from the photon to the electron to eject the electron from the metal. The photon must have enough energy to overcome the work function. An electron ejected from a metal because of a collision with a photon is called a photoelectron.

Photovoltaics

Photovoltaics is an application of the photoelectric effect. We can describe photovoltaics as the use of light to generate electrical current. We can make a photovoltaic cell, or photo-cell, by placing two semi-conductors in contact with each other. Light shining on the photocell expels electrons that can travel from one semi-conductor to the other.

Photocells are not sources of energy and they do not store energy. Photocells transform sunlight into electrical energy. When sunlight is removed, the photocell will stop producing electricity. If a photocell is used to produce electricity, an additional system is also needed to provide energy when light is not available. The additional system can be an energy storage system that is charged by sunlight transformed into electrical energy and then stored, or it can be a supplemental energy supply provided by another source.

Point to Ponder: What are some current uses of photo-voltaics?

There are many applications of photovoltaics. They include solar powered calculators, solar powered camping equip-ment, solar powered satellites, the International Space Station, and solar powered cars. A solar powered traffic system for helping enforce speed limit laws is shown in Figure 4-11.

Figure 4-11. Solar Powered Traffic Control, Golden, Colorado

Chapter 5

ENERGY OPTIONS – WIND AND WATER

      Fossil energy, nuclear energy, and solar energy provide most of the energy used in the world today and are expected to contribute to the global energy mix for decades to come. Our purpose here is to introduce a range of non-solar renewable energy sources. Renewable energy was defined in Chapter 4 as energy obtained from sources at a rate that is less than or equal to the rate at which the source is replenished. We consider renewable energy technology below that is based on wind and water.

WIND

Wind energy technology relies on gradients in physical properties such as atmospheric pressure to generate electrical power. The kinetic energy of wind and flowing water are indirect forms of solar energy and are therefore considered renewable. Wind turbines harness wind energy and hydroe-lectric energy is generated by the flow of water through a turbine. Both convert the mechanical energy of a rotating blade into electrical energy in a generator.

Wind has been used as an energy source for thou-sands of years. Historical applications include sailing and driving windmills. Windmills have been used for grinding grain and pumping water. Wind is still used today as a source of power for sailing vessels and parasailing. The use of wind as a source of energy for generating electrical power is a relatively new but rapidly growing technology. It is the primary focus of this section.

Wind Turbine

Modern wind turbines are classified as either horizontal axis turbines or vertical axis turbines. A vertical axis turbine has blades that rotate around a vertical axis and its visual ap-pearance has been likened to an eggbeater. A horizontal axis turbine has blades that rotate around a horizontal axis (see Figure 5-1). Horizontal axis turbines are the most common turbines in use today.

A typical horizontal axis turbine consists of a rotor with two or more blades attached to a machine cabin set atop a post that is mounted on a foundation block. The machine cabin contains a generator attached to the wind turbine. The rotor blades can rotate in the vertical plane and the machine cabin can rotate in the horizontal plane. If the speed of rotation of the tip of the rotor blade is fast enough, it can be lethal to birds entering the fan area of the rotor blade. This environmental hazard can be minimized by selecting locations for wind turbines that avoid migration patterns. Another way to minimize environmental impact is to use large rotor blades that turn at relatively low speeds of rotation.

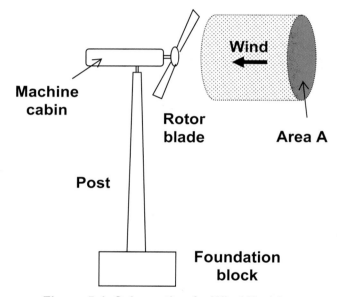

Figure 5-1. Schematic of a Wind Turbine

Modern wind turbines are several hundred feet tall. For example, a wind turbine that generates 1.6 megawatts of electrical power is approximately 113 meters (370 feet) tall from its base to the tip of the rotor blade. The oil storage tanks in Figure 5-2 help illustrate the scale of a modern wind turbine and demonstrate that wind turbines can be erected on existing industrial properties. We have seen wind turbines erected near fossil fuel fired power plants, along highways, and on ranchland.

Figure 5-2. Modern Wind Turbine near Rotterdam, Holland

Wind power is at maximum if the wind direction is perpendicular to the plane of rotation of the rotor blade. If the wind direction is parallel to the plane of rotation of the rotor blade for an infinitesimally thin rotor blade, the wind turbine will not provide any wind power. A change in wind direction can put stress on wind turbines. In addition, wind speed is seldom constant; it can vary from still to tornado or hurricane speed.

One factor that affects the efficiency of a wind turbine is the efficiency of converting mechanical energy of the rotor blade into electrical energy. In 1928, Albert Betz showed that the maximum percentage of wind power that can be extracted is approximately 59.3% of the power in the wind. Another factor affecting electrical power output from a wind turbine is the reliability of the wind turbine. The rate of rotation of the rotor blade depends on wind speed. If the wind speed is too large, the rotor blade can turn too fast and damage the system. To avoid this problem, wind turbines may have to be taken off-line in high wind conditions, which reduces their reliability.

Wind Farms

A wind farm or wind park is a collection of wind turbines. The areal extent of the wind farm depends on the radius R of the rotor blades (Figure 5-3). A wind turbine must have enough space around the post to allow the fan of the rotor blade to face in any direction. The minimum spacing between the posts of two equivalent wind turbines must be $2R_{eff}$ to avoid collisions between rotor blades. If we consider the aerodynamics of wind flow, which is the factor that controls turbine spacing, the turbine spacing in a wind farm increases to at least 5 to 10 times rotor diameter $2R$ [Sørensen, 2000, page 435] behind the plane of the rotor blade. The additional distance between posts is designed to minimize turbulence between wind turbines and enable the restoration of the wind stream to its original undisturbed state after it passes by one turbine on its way to the next turbine. Wind turbine spacing is an important factor in determining the surface area, or footprint, needed by a wind farm.

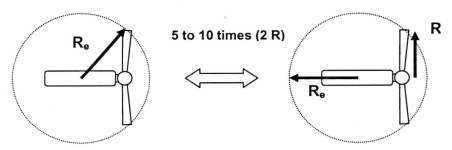

Figure 5-3. Wind Turbine Spacing

Environmental Impact

Wind energy is a renewable energy that is considered a clean energy because it has a minimal impact on the environment compared to other forms of energy. Wind turbines provide electrical energy without emitting greenhouse gases. On the other hand, we have already observed that the harvesting of wind energy by wind turbines can have environmental consequences.

Rotating wind turbine blades can kill birds and interfere with migration patterns. Wind turbines with slowly rotating, large diameter blades and judicious placement of wind turbines away from migration patterns can reduce the risk to birds. Wind farms can have a significant visual impact that may be distasteful to some people. Wind turbines produce some noise when they operate. In the past, wind turbines with metal blades could interfere with television and radio signals. Today, turbine blades are made out of composite materials that do not interfere with electromagnetic transmissions.

Point to Ponder: Can wind provide all of our energy needs?

Wind appears to have many of the advantages that would make it an appealing solution to our energy problems. When properly designed and located, wind turbines are environmentally benign. Some people may object to large fields filled with wind turbines, but fields of wind turbines can be built on property to serve a dual purpose. For example, wind farms can be built on West Texas ranch land or along the roadways in Rotterdam, Holland. In both cases the turbines rotate well above activities below. Modern turbines rotate at a slow enough speed that birds can see them and avoid them. The question is, how many wind turbines would we need to supply global energy demand?

If we assume the world population in year 2100 will be eight billion people and the amount of energy needed to provide each person an acceptable quality of life will be 200,000 megajoules per year, we can estimate the number of wind turbines we would need. Suppose we use wind turbines that can provide four megawatts each. We would need about 12.7 million wind turbines to supply global energy demand in 2100. If these wind turbines are collected in wind farms that can provide 1000 megawatts per wind farm, we would need approximately 50,700 wind farms [Fanchi, 2004, Exercises 15-7 and 12-8]. If we assume the turbine radius is 108 feet and assume the area occupied by each wind farm is approximately square, with a turbine separation of about ten times turbine radius, we estimate that each wind farm will occupy about nine square miles. All of the farms would occupy an area of about 465,000 square miles, or about 16% of the area in the continental United States. The area is smaller than the state of Alaska, but larger than the state of Texas.

It is clear that we could build enough wind farms to provide the energy we need based on area and power capacity. Other issues must also be considered. We need to provide energy when the wind does not blow. We need to distribute the energy where it is needed and when it is needed. We need to provide energy in a form that best fits the need. We need to be willing to accept the environmental impact of wind farms, including their appearance and impact on wildlife. Some of these unresolved issues are technical, others are social. An example of a technical issue is the question of how to provide energy on demand, even when the wind is not blowing. There are several options to consider. For example, we could use wind energy to charge batteries or we could use wind energy to produce hydrogen for use in fuel cells. One advantage of hydrogen production is that we could use hydrogen in the transportation sector.

Much work remains to be done to identify the optimum strategy for providing renewable energy. It is important to note, however, that by late 2003, wind power in the United States cost approximately US$0.05 per kilowatt hour, which was comparable to the cost of electricity from natural gas. The cost-competitiveness of wind and the demand for cleaner energy by consumers is encouraging the growth of wind energy around the world.

HYDROPOWER

People have known for some time that falling water could be used to generate electric power. Many of the first commercial electric power plants relied on flowing water as their primary energy source. A schematic of a hydroelectric power plant is presented in Figure 5-4.

Water flows from a higher elevation in Figure 5-4 to a lower elevation through a pipeline called a penstock. The change in elevation of the water is a change in the potential energy of the water. Dams with turbines and generators convert the change in potential energy into mechanical kinetic energy. The water current turns a turbine that is connected to a generator. The turbine rotates the generator shaft. The generator shaft is connected to either a magnet adjacent to a coil of wire or a coil of wire adjacent to a magnet. The resulting alternating current generator converts the mechanical energy of rotation to electrical energy.

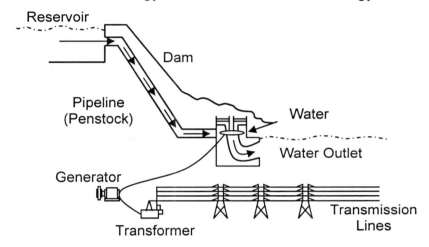

Figure 5-4. Principles of Hydroelectric Power Generation

The elevation, or head, of the hydroelectric power plant is the height the water falls. The head is replaced by the effective head for realistic systems. The effective head is less than the actual head, or elevation, because water flowing through a conduit such as a pipe will lose energy to friction and turbulence. The rate that water falls through the effective head depends on the volume of the penstock shown in Figure 5-4. If the penstock volume is too small, the output power will be less than optimum because the flow rate could have been larger. On the other hand, the penstock volume cannot be arbitrarily large because the flow rate through the penstock depends on the rate that water fills the reservoir behind the dam. Figure 5-5 shows the reservoir behind the Hoover Dam at the Arizona-Nevada border in the

United States. The light colored rock in the figure shows that the depth of the reservoir is below normal because of an extended drought at the time the photo was taken.

Figure 5-5. Reservoir behind Hoover Dam

The volume of water in the reservoir and the corresponding water level depends on the water flow rate into the reservoir. During drought conditions, the elevation of the water level can decline because there is less water in the reservoir. During rainy seasons, the elevation of the water level can increase as more water drains into the streams and rivers that flow into the reservoir behind the dam. Hydropower facilities must be designed to balance the flow of water through the electric power generator with the water that fills the reservoir through such natural sources as rainfall, snowfall, and drainage. Some typical hydropower plant sizes are shown in Table 5-1.

Table 5-1. Hydropower Plant Sizes [DoE Hydropower, 2002]	
Size	**Electrical Energy Generating Capacity (MW)**
Micro	< 0.1
Small	0.1 to 30
Large	> 30

WAVES AND TIDES

The oceans are another solar-powered source of energy. The mechanical energy associated with waves and tides, and thermal energy associated with temperature gradients in the ocean can be used to drive electric generators. We consider each of these energy sources in this section.

Waves and Tides

In our discussion of hydropower, we saw that water can generate power when it moves from a high potential energy state to a low potential energy state, like water flowing over a waterfall. If the moving water is an ocean wave, the elevation varies sinusoidally with time. The sinusoidal wave has crests and troughs. The wave amplitude, or height of the wave relative to a calm sea, depends on weather conditions: it will be small during calm weather and can be very large during inclement weather such as hurricanes. The build-up of an ocean wave off the coast of Oahu is shown in Figure 5-6. The change in the potential energy of wave motion can be transformed into energy for performing useful work.

Figure 5-6. Motion of an ocean wave, Oahu, Hawaii

Suppose we lay a paddle on the water. The paddle should be buoyant enough to move up and down with the wave (Figure 5-7). The change in potential energy of the paddle depends on the amplitude of the water wave, the size of the paddle, and the density of the paddle.

The energy density of waves breaking on a coastline in favorable locations can average 65 megawatts/mile (40 megawatts/kilometer) of coastline [DoE Ocean, 2002]. The motion of the wave can be converted to mechanical energy and used to drive an electricity generator. The wave energy can be captured by floats or pitching devices like the paddles presented in Figure 5-7. Another approach to capturing wave energy is to install an oscillating water column. The rise and fall of water in a cylindrical shaft drives air in and out of the top of the shaft. The motion of the air provides power to an air-driven turbine. The output power depends on the efficiency of converting wave motion to mechanical energy and then converting mechanical energy to electrical energy. A third approach to capturing wave energy is the wave surge or focusing technique. A tapered channel installed on a shoreline concentrates the water waves and channels them into an elevated reservoir. Water flow out of the reservoir is combined with hydropower technology to generate electricity.

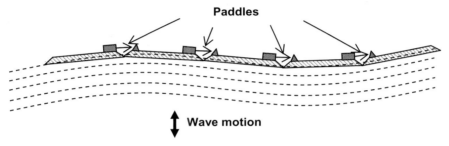

Figure 5-7. Capturing Wave Energy

The ebb and flow of tides produces tidal energy that can be captured to produce electricity. Figure 5-8 illustrates a tidal energy station. A dam with a sluice is erected across the opening to a tidal basin to capture tidal energy. The sluice is opened to let the tide flow into the basin and then closed when sea level drops. The water can flow through the gate four times per day because the tide rises and falls twice a day. Hydropower technology can be used to generate electricity from the elevated water in the basin.

Tides can rise as high as 15 meters on the Rance River in France. The Rance tidal energy station has the potential of generating 240 megawatts of power [DoE Ocean, 2002; Serway and Faughn, 1985, page 114]. Tidal energy stations can have an environmental impact on the ecology of the tidal basin because of reduced tidal flow and silt buildup.

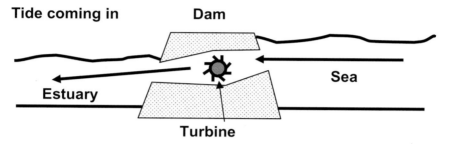

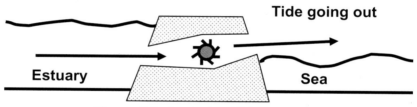

Figure 5-8. Capturing Tidal Energy

Point to Ponder: How much energy can be obtained from coastal waves?

A typical fossil fuel fired power plant produces 1000 megawatts of power. If we assume that the average wave power along a coast is 30 megawatts per kilometer and assume the efficiency of power production is 25%, we would need to produce energy from 133 kilometers (82 miles) of coastline to provide 1000 megawatts of power. [Fanchi, 2004, Exercise12-6]

Ocean Thermal

Temperature gradients in the ocean exist between warm surface water and cooler water below the surface. Near ocean bottom geothermal vents and underwater volcanoes, temperature gradients exist between hot water near the heat source and cooler waters away from the heat source. If the temperature gradients are large enough, they can be used to generate power using ocean thermal energy conversion (OTEC) power plants.

Three types of OTEC systems are recognized [DoE Ocean, 2002]: closed-cycle plants, open-cycle plants, and hybrid plants. A closed-cycle plant circulates a heat transfer fluid through a closed system. The fluid is heated with warm seawater until it is flashed to the vapor phase. The vapor is routed through a turbine and then condensed back to the liquid phase with cooler seawater. Pressure changes in an open-cycle plant make it possible to flash warm seawater to steam and then use the steam to turn a turbine. A hybrid plant is a combination of an open-cycle plant with a heat transfer fluid. Warm seawater is flashed to steam in a hybrid plant and then used to vaporize the heat transfer fluid. The vaporized heat transfer fluid circulates in a closed system as in the closed-cycle plant.

An ocean thermal energy conversion system can be built on land near a coast, installed near the shore on a continental shelf, or mounted on floating structures for use offshore. In some parts of the world, a desirable byproduct of an OTEC system is the production of desalinated water. Salt from the desalination process must be disposed of in a manner that minimizes its environmental impact.

GEOTHERMAL

Geothermal energy is energy provided by the Earth. The Earth's interior is subdivided into a crystalline inner core, molten outer core, mantle, and crust. Basalt, a dark volcanic rock, exists in a semi-molten state at the surface of the mantle just beneath the crust. Drilling in the Earth's crust has shown that the temperature of the crust tends to increase linearly with depth. The interior of the Earth is much hotter than the crust. The source of heat energy is radioactive decay, and the crust of the Earth acts as a thermal insulator to prevent heat from escaping into space.

Geothermal energy is obtained by using a fluid (such as hot water or steam) to carry the internal energy of the Earth to the surface of the Earth. Geothermal energy can be obtained from temperature gradients between the shallow ground and surface, subsurface hot water, hot rock several kilometers below the Earth's surface, and magma. Magma is molten rock in the mantle and crust that is heated by the large heat reservoir in the interior of the Earth. In some parts of the crust, magma is close enough to the surface of the Earth to heat rock or water in the pore spaces of rock. The heat energy acquired from geological sources is called geothermal energy. Magma, hot water, and steam are carriers of energy. The energy is displayed when lava flows into the sea and creates steam as shown in Figures 5-9 and

5-10. In addition to steam, the cloud contains toxic chemicals such as hydrogen sulfide and even glass particles.

Figure 5-9. Steam caused by lava meeting the Pacific, Hawaii

Some of the largest geothermal production facilities in the world are at the Geysers in California, and in Iceland. The location of these areas is determined by the proximity of a suitable geothermal energy source. For example, the Puna geothermal power plant on the Big Island of Hawaii is situated south of Hilo near Volcano National Park.

The technology for converting geothermal energy into useful heat and electricity can be categorized as geothermal heat pumps, direct-use applications, and geothermal power plants. Each of these technologies is discussed below.

Figure 5-10. Lava from the Pu'u O'o vent, Kilauea Volcano, Hawaii

Geothermal Heat Pump

A geothermal heat pump uses energy near the surface of the Earth to heat and cool buildings. The temperature of the upper three meters (10 feet) of the Earth's crust remains in the relatively constant range of 10° Centigrade to 16° Centigrade. A geothermal heat pump for a building consists of ductwork in the building connected through a heat exchanger to pipes buried in the shallow ground nearby. The building can be heated during the winter by pumping water through the geothermal heat pump. The water is warmed when it passes through the pipes in the ground. The resulting heat is carried to the heat exchanger where it is used to warm air in the ductwork. During the summer, the direction of heat flow is reversed. The heat exchanger uses heat from hot air in the building to warm water that carries the heat through the pipe system into the cooler shallow ground. In the winter, heat is added to the building from the Earth, and in the summer heat is removed from the building.

Direct-Use Applications

A direct-use application of geothermal energy uses heat from a geothermal source directly in an application. Hot water from the geothermal reservoir is used without an intermediate step such as the heat exchanger in the geothermal heat pump. Hot water from a geothermal reservoir may be piped directly into a facility and used as a heating source. A direct-use application for a city in a cold climate with access to a geothermal reservoir is to pipe the hot water from the geothermal reservoir under roads and sidewalks to melt snow.

Minerals that are present in the geothermal water will be transported with the hot water into the pipe system of the direct-use application. Some of the minerals will precipitate out of the water when the temperature of the water decreases. The precipitate will form a scale in the pipes and reduce the flow capacity of the pipes. Filtering the hot water or adding a scale retardant can reduce the effect of scale. In either case, the operating costs will increase.

Geothermal Heating System

An example of a geothermal application with a heat exchanger is the geothermal heating system sketched in Figure 5-11. The heat exchanger is shown at the top of the figure and the geothermal reservoir is shown at the base of the figure.

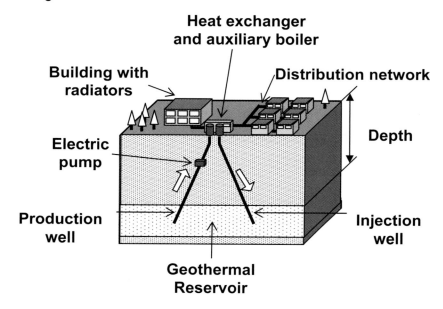

Figure 5-11. Geothermal Heating System

A geothermal reservoir is an aquifer with hot water or steam. The production well is used to withdraw hot water from the geothermal reservoir and the injection well is used to recycle the water. Recycling helps maintain reservoir pressure. If the geothermal reservoir is relatively small, the recycled, cooler water can lower the temperature of the aquifer. The electric pump in the figure is needed to help withdraw water because the reservoir pressure in this case is not high enough to push the water to the surface. Heat from the geothermal reservoir passes through a heat exchanger and is routed to a distribution network.

Geothermal Power Plants

Geothermal power plants use steam or hot water from geothermal reservoirs to turn turbines and generate electricity. Dry steam power plants use steam directly from a geothermal reservoir to turn turbines. Flash steam power plants allow higher-pressure hot water from a geothermal reservoir to flash to steam in lower-pressure tanks. The resulting steam is used to turn turbines. A third type of geothermal power plant called a binary-cycle plant uses heat from moderately hot geothermal water to flash a second fluid to the vapor phase. The second fluid must have a lower boiling point than water so that it will be vaporized at the lower temperature associated with the moderately hot geothermal water. There must be enough heat in the geothermal water to supply the latent heat of vaporization needed by the secondary fluid to make the phase change from liquid to vapor. The vaporized secondary fluid is then used to turn turbines.

Managing Geothermal Reservoirs

Geothermal reservoirs rely on production and injection of subsurface fluids through wells. Figure 5-12 shows the rig used to drill an injection well at Puna Geothermal Power Plant on the Big Island of Hawaii.

Figure 5-12. Drilling Rig at the Puna Geothermal Power Plant, Big Island of Hawaii

Like oil and gas reservoirs, the hot water or steam in a geothermal reservoir can be depleted by production. The phase of the water in a geothermal reservoir depends on the pressure and temperature of the reservoir. Single-phase steam will be found in low pressure, high temperature reservoirs. In high-pressure reservoirs, the water may exist in the liquid phase, or in both the liquid and gas phases, depending on the temperature of the reservoir.

When water is produced from the geothermal reservoir, both the pressure and temperature in the reservoir can decline. In this sense, geothermal energy is a nonrenewable, finite resource unless the produced hot water or steam is replaced. A new supply of water can be used to replace the produced fluid or the produced fluid can be recycled after heat transfer at the surface. If the rate of heat transfer from the heat reservoir to the geothermal reservoir is slower than the rate of heat extracted from the geothermal reservoir, the temperature in the geothermal reservoir will decline during production. To optimize the performance of the geothermal reservoir, it must be understood and managed in much the same way that petroleum reservoirs are managed.

Hot Dry Rock

Another source of geothermal energy is hot, dry rock several kilometers deep inside the Earth. These rocks are heated by magma directly below them and have elevated temperatures, but they do not have a means of transporting the heat to the surface. In this case, it is technically possible to inject water into the rock, let it heat up and then produce the hot water. Figure 5-13 illustrates a hot, dry rock facility.

The facility is designed to recycle the energy carrying fluid. Water is injected into fissures in the hot, dry rock through the injector and then produced through the producer. The power plant at the surface uses the produced heat energy to drive turbines in a generator. After the hot, produced fluid transfers its heat to the power plant, the cooler fluid can be injected again into the hot, dry rock.

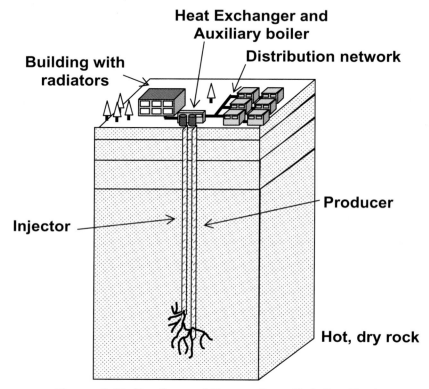

Figure 5-13. Geothermal Energy from Hot, Dry Rock

Point to Ponder: Can we use water to provide our energy needs?

Water is available worldwide and has been used to provide energy for centuries. It does have limitations, however. Hydropower is already widely used and can impact the environment. Geothermal energy can be a commercial source of energy in the locations where it is accessible. Waves and tides can provide commercial energy, but the efficiency of energy production is relatively low and the environmental impact, especially the visual impact along coasts, can be significant.

Although water appears to have a limited ability to satisfy our energy needs, water does contain an element that has much greater potential: hydrogen. We discuss the role that hydrogen may play in the future energy mix in Chapter 7.

Chapter 6

ENERGY OPTIONS – BIOMASS AND SYNFUELS

Biomass refers to wood and other plant or animal matter that can be burned directly or can be converted into fuels. Wood has historically been a source of fuel. Now, technologies exist to convert plants, garbage, and animal dung into natural gas. Synthetic fuels (synfuels) are fossil fuel substitutes created by chemical reactions using such basic resources as coal or biomass. Synthetic fuels are used as substitutes for conventional fossil fuels such as natural gas and oil. We discuss biomass and synfuels in this chapter.

BIOMASS

People have used biomass for fuel ever since we learned to burn wood. Biomass is matter that was recently formed as a result of photosynthesis. Biomass includes wood and other plant or animal matter that can be burned directly or can be converted into fuels. The compost heap shown in Figure 6-1 is an example of biomass. In addition, products derived from biological material are considered biomass. Methanol, or wood alcohol, is a volatile fuel that has been used in race cars for years. Another alcohol, clean burning ethanol, can be blended with gasoline to form a blended fuel (gasohol) and used in conventional automobile engines, or used as the sole fuel source for modified engines.

Figure 6-1. Compost Heap at Rudy's Peach Orchard, Spring, Texas

Availability is one advantage biomass has relative to other forms of renewable energy. Energy is stored in biomass until it is needed. The availability of other renewable energy forms, such as wind and solar energy, depend on environmental conditions that can vary considerably.

Biomass can be used as a fuel in the solid, liquid, or gaseous state. Technologies exist to convert plants, garbage (Figure 6-1), and animal dung into natural gas. An example of a biomass project is the production of gas from a landfill. A landfill is a pit filled with garbage. When the pit is full, we can cover it with dirt and drill a well through the dirt into the pit. The well provides a conduit for producing the natural gas that is generated from the decay of biological waste in garbage. The landfill gas is filtered, compressed, and routed to the main gas line for delivery to consumers.

The amount of energy that can be produced from biomass depends on the heat content of the material when it is dry. Table 6-1 shows energy densities for several common materials. A kilogram (2.2 pounds mass) of dry wood has a smaller energy density than crude oil and coal. The energy density of biomass is less than the energy density of fossil fuel. Consequently, the combustion of fossil fuels can provide more heat than the combustion of biomass. Natural gas burners, for example, are more efficient than boilers used with wood or straw [Sørensen, 2000, page 473].

Table 6-1. Energy Density of Common Materials		
Material	MJ kg^{-1}	MJ m^{-3}
Crude oil[a]	42	37,000
Coal[a]	32	42,000

Table 6-1. Energy Density of Common Materials		
Material	MJ kg^{-1}	MJ m^{-3}
Dry wood or sawmill scrap[b]	12.5	10,000
Methanol[a]	21	17,000
Ethanol[a]	28	22,000
[a]Sørensen [2000, page 552] [b]Sørensen [2000, page 473]		

Some types of biomass used for energy include forest debris, agricultural waste, wood waste, animal manure, and the non-hazardous, organic portion of municipal solid waste. Developing countries are among the leading consumers of biomass because of their rapid economic growth and increasing demand for electricity in municipalities and rural areas.

Point to Ponder: Why is energy density important?
Energy density tells us how much energy we can obtain from a given amount of material and it is a quantity we can use to compare different fuels. For example, one kilogram (2.2 pounds mass) of crude oil provides 42 megajoules of energy, while one kilogram of sawmill scrap provides 12.5 megajoules of energy. Thus, a kilogram of crude oil provides over three times as much energy during the combustion process than the same mass of sawmill scrap.

Wood

Wood has historically been a source of fuel. We have already learned that wood was a primary energy source through much of history [Nef, 1977], but deforestation became such a significant problem in 16th century England that the English sought, and found, an alternative source of fuel: coal. Today, many people rely on wood as a fuel source in underdeveloped parts of the world. Economic sources of wood fuels include wood residue from manufacturers, discarded wood products, and non-hazardous wood debris.

An increased reliance on wood as a fuel has environmental consequences, such as pollution, an increased rate of deforestation, and an increase in the production of byproducts from wood burning. Figure 6-2 shows the afternoon Sun as it was seen through smoke from a forest fire. One way to mitigate the environmental impact associated with burning wood is to implement reforestation and carbon sequestration programs. Each of these programs is discussed below.

Reforestation programs are designed to replenish the forests. Research is underway to genetically engineer trees that grow fast, are drought resistant, and easy to harvest. Fast-growing trees are an example of an energy crop; that is, a crop that is genetically designed to become a competitively priced fuel.

We know from observation and the chemistry of wood combustion that burning wood consumes carbon, hydrogen and oxygen. The combustion reactions produce water, carbon monoxide, and carbon dioxide, a greenhouse gas. Efforts to reduce greenhouse gas emissions, such as sequestering (or storing) greenhouse gases in geologic formations, are as important to biomass consumption as they are to fossil fuel consumption.

Figure 6-2. Afternoon Sun Obscured by Smoke from a Forest Fire near Denver, Colorado

Ethanol

Ethanol is an alcohol that can be added to gasoline to increase the volume of fuel available for use in internal combustion engines. Ethanol is made from a biomass feedstock. A feedstock is the raw material supplied to an industrial processor. Residues that are the organic byproducts of food, fiber, and forest production are economical biomass feedstock. Examples of residue include sawdust, rice husks, corn stover, and bagasse. Corn stover is used to produce ethanol. It is a combination of corn stalks, leaves, cobs, and husks. Bagasse is the residue that is left after juice has been extracted from sugar cane.

The process of ethanol production proceeds in several steps. First, the feedstock must be delivered to the feed handling area. The biomass is then conveyed to a pretreatment area where it is soaked for a short period of time in a dilute sulfuric acid catalyst at a high temperature. The pretreatment process liberates certain types of sugars and other compounds. The liquid portion of the mixture is separated from the solid sludge and washed with a base such as lime to neutralize the acid and remove compounds that may be toxic to fermenting organisms. A beer containing ethanol is produced after several days of anaerobic fermentation. The beer is distilled to separate ethanol from water and residual solids.

The production of ethanol from corn requires distillation. During distillation, a mixture is boiled and the ethanol-rich vapor is allowed to condense to form a sufficiently pure alcohol. Distillation is an energy consuming process. The use of ethanol in petrofuels to increase the volume of fuel would yield a favorable energy balance if the energy required to produce the ethanol were less than the energy provided by the ethanol during the combustion process. Researchers are attempting to determine if ethanol use serves as a source of energy or as a net consumer of energy. A biomass research facility at the National Renewable Energy Laboratory in Golden, Colorado is shown in Figure 6-3. A fermentation tank is shown in the foreground of the figure.

Figure 6-3. Biomass Research Facility, National Renewable Energy Laboratory, Golden, Colorado

Biopower

Biomass has historically been used to provide heat for cooking and comfort. Today, biomass fuels can be used to generate electricity and produce natural gas. The power obtained from biomass, called biopower, can be provided in scales ranging from single-family homes to small cities.

Biopower is typically supplied by one of four different classes of systems: direct-fired, cofired, gasification, and modular systems.

A direct-fired system is similar to a fossil fuel fired power plant. High-pressure steam for driving a turbine in a generator is obtained by burning biomass fuel in a boiler. Biopower boilers presently have a smaller capacity than coal-fired plants. Biomass boilers provide energy in the range of 20 to 50 megawatts, while coal-fired plants provide energy in the range of 100 to 1500 megawatts. The technology exists for generating steam from biomass with an efficiency of over 40%, but actual plant efficiencies tend to be on the order of 20% to 25%.

Biomass may be combined with coal and burned in an existing coal-fired power plant. The process is called cofiring and is considered desirable because the combustion of the biomass-coal mixture is cleaner than burning only coal. Biomass combustion emits smaller amounts of pollutants, such as sulfur dioxide and nitrogen oxides. The efficiency of the cofiring system is comparable to the coal-fired power plant efficiency when the boiler is adjusted for optimum performance. Existing coal-fired power plants can be converted to cofiring plants without major modifications. The efficiency of transforming biomass energy to electricity is comparable to the efficiency of a coal-fired power plant and is in the range of 33% to 37%. The use of biomass that is less expensive than coal can reduce the cost of operating a coal-fired power plant.

Biomass gasification converts solid biomass to flammable gas. Cleaning and filtering can remove undesirable compounds in the resulting biogas. The gas produced by gasification can be burned in combined-cycle power generation systems. The combined-cycle system combines gas turbines and steam turbines to produce electricity with efficiency as high as 60%.

Modular systems use the technologies described above, but on a scale that is suitable for applications that demand less energy. A modular system can be used to supply electricity to villages, rural sites, and small industry.

Point to Ponder: Is biomass a desirable alternative energy source?

Figure 6-4 shows a car that uses 100% vegetable oil as its fuel. At the time the photo was taken, a gallon of gasoline in Honolulu, Hawaii cost approximately one third more than a gallon of gasoline in Denver, Colorado. Increases in the price of gasoline stimulate the search for alternative transportation fuels. Vegetable oil is an example of a biomass alternative energy source.

Biomass provides energy by generating heat using combustion. One byproduct of the combustion process is carbon dioxide, a greenhouse gas. Thus, even though biomass is renewable and can serve as an energy source, the combustion byproduct carbon dioxide must be prevented from accumulating in the atmosphere. In this sense, biomass is no more desirable than fossil fuels. The need to sequester carbon dioxide is a technical and social issue that is discussed further in Chapter 9.

Figure 6-4. Car Powered by 100% Vegetable Oil, Waikiki, Oahu

SYNFUELS

Synthetic fuels, or synfuels, are fossil fuel substitutes cre-ated by chemical reactions using such basic resources as coal or biomass. There are several ways to convert biomass into synfuels. Oils produced by plants such as rapeseed (canola), sunflowers and soybeans can be extracted and refined into a synthetic diesel fuel that can be burned in diesel engines. Thermal pyrolysis and a series of catalytic

reactions can convert the hydrocarbons in wood and municipal wastes into a synthetic gasoline.

Society can extend the availability of fossil fuels such
as natural gas and oil by substituting synthetic fuels for
conventional fossil fuels. Here we consider coal gasification,
biomass conversion, and gas-to-liquids conversion.

Coal Gasification

Large coal molecules are converted to gaseous fuels in coal
gasification. Coal is thermally decomposed at temperatures
on the order of 600° Centigrade (1112° Fahrenheit) to 800°
Centigrade (1472° Fahrenheit). The products of decomposition are methane and a carbon rich char. If steam is present,
the char will react with the steam in the reaction to form
carbon monoxide and hydrogen. The carbon monoxide-
hydrogen mixture is called synthesis gas, or syngas. Carbon
monoxide can react with steam in the reaction to form carbon dioxide and hydrogen. If the carbon dioxide is removed,
the hydrogen-enriched mixture will react with carbon monoxide to produce methane and water vapor. The coal
gasification process can be used to synthesize methane
from coal. Methane is easier to transport than coal.

Biomass Conversion

Biomass, particularly biological waste, appears to be a
plentiful source of methane. A simple methane digester that
converts biological feed such as dung to methane is shown
in Figure 6-5. The screw agitator mixes the liquid slurry

containing the feed. The mixing action facilitates the release of methane from the decaying biomass as a biomass conversion process. Methane trapped by the metal dome is recovered through the gas outlet.

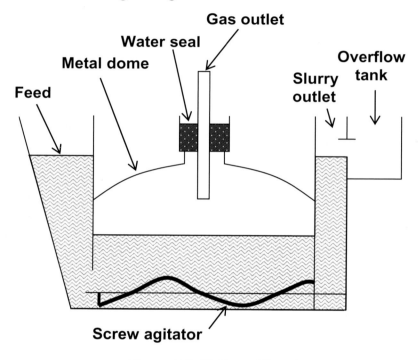

Figure 6-5. Methane Digester

Fermentation is another example of a biomass conversion process. Microorganisms are used in the fermentation process to convert fresh biological material into simple hydrocarbons or hydrogen. We illustrate fermentation processes by describing the anaerobic digestion process, a process that is well suited for producing methane from biomass.

The anaerobic digestion process proceeds in three stages. In the first stage, the complex biomass is decomposed by the first set of microorganisms. The decomposition of cellulosic material into the simple sugar glucose occurs in the presence of enzymes provided by the microorganisms. Stage one does not require an anaerobic (oxygen-free) environment. In the second stage, hydrogen atoms are removed in a dehydrogenation process that requires acidophilic (acid-forming) bacteria. In the third stage, a mixture of carbon dioxide and methane called biogas is produced from the acetic acid produced in stage two. The third stage requires the presence of anaerobic bacteria known as methanogenic bacteria in an oxygen-free environment.

Biomass conversion turns a waste product into a useful commodity. One difficulty with the exploitation of biomass fuels is the potential impact on the ecology of the region. For example, excessive use of dung and crop residues for fuel instead of fertilizer can deprive the soil of essential nutrients that are needed for future crops. Proper crop management can minimize this concern.

Gas-to-Liquid Conversion
Synthetic liquid hydrocarbon fuels can be produced from natural gas by a gas-to-liquids (GTL) conversion process. The primary product of the GTL process is a low sulfur, low aromatic diesel fuel.

Point to Ponder: Are synfuels desirable alternative energy sources?

Like biomass and fossil fuels, synfuels provide energy by generating heat using combustion. Once again the combustion process produces carbon dioxide as a byproduct. Carbon dioxide is a greenhouse gas and sequestration becomes an issue. Sequestration is the capture and storing of greenhouse gases. For example, we can geologically sequester carbon dioxide by injecting it into a geologic formation such as a coal seam or an abandoned oil field. Unlike biomass, coal gasification and gas-to-liquid conversion depend on the fossil fuels coal and natural gas. These synfuels are only renewable to the extent that coal and natural gas are renewable. In the case of biomass conversion, natural gas can be renewed from biological materials such as organic wastes, but a byproduct is carbon dioxide, an undesirable greenhouse gas. The rate of generation of natural gas from biomass conversion will limit the amount of energy that can be provided by biomass and synfuels.

Chapter 7

HYDROGEN – AN ENERGY CARRIER

Hydrogen is found almost everywhere on the surface of the Earth as a component of water (Figure 7-1). It has many commercial uses, including ammonia production for use in fertilizers, methanol production, hydrochloric acid production, and use as a rocket fuel. Liquid hydrogen is used in cryogenics and superconductivity. Hydrogen is important to us because it can be used as a fuel.

Hydrogen can be used as a fuel for a modified internal combustion engine or in a fuel cell. Fuel cells are electrochemical devices that directly convert hydrogen, or hydrogen-rich fuels, into electricity using a chemical process. Fuel cells do not need recharging or replacing and can produce electricity as long as they are supplied with hydrogen and oxygen. Hydrogen can be produced by the electrolysis of water, which uses electrical energy to split water into its constituent elements. Electrolysis is a net energy consuming process; electrolysis consumes more energy than it produces. Hydrogen can also be produced from certain types of algae. Hydrogen is not considered an energy source because more energy is required to produce hydrogen than can be obtained from hydrogen. Hydrogen is considered a carrier of energy because it can be used as a fuel to provide energy. The properties of hydrogen make it a source of hope for a sustainable, global economy. We first describe the properties of hydrogen before discussing the production of hydrogen and its use as the primary fuel in fuel cells.

Figure 7-1. Each water molecule in this mountain reservoir in Summit County, Colorado contains two atoms of hydrogen and one atom of oxygen.

PROPERTIES OF HYDROGEN

Hydrogen is the first element in the Periodic Table. The nucleus of hydrogen is the proton and hydrogen has only one electron. Selected physical properties of hydrogen, methane and gasoline are shown in Table 7-1.

Table 7-1. Selected Physical Properties of Hydrogen, Methane, and Gasoline [a]			
Property	Hydrogen (gas)	Methane (gas)	Gasoline (liquid)
Molecular Weight (grams/moles) [a]	2.016	16.04	~110
Mass Density (kilograms/meters^3) [a,b]	0.09	0.72	720-780
Energy Density (megajoules/kilograms)	120 [a]	53 [c,d]	46 [a,c]
Volumetric Energy Density (megajoules/meters^3) [a]	11 [a]	38 [c,d]	35,000 [a,c]
Higher Heating Value (megajoules/kilograms) [a]	142.0	55.5	47.3
Lower Heating Value (megajoules/kilograms) [a]	120.0	50.0	44.0

a. Ogden [2002, Box 2, page 71]
b. at 1 atm and 0 °C
c. Hayden [2001, page 183]
d. Ramage and Scurlock [1996, Box 4.8, page 152]

At ambient conditions on Earth, hydrogen is a colorless, odorless, tasteless, and non-toxic gas of diatomic molecules. A kilogram (2.2 pounds mass) of hydrogen, in either the gas or liquid state, has a greater energy density than the most widely used fuels today, such as oil and coal.

Point to Ponder: Is hydrogen the perfect fuel of the future?

Hydrogen is considered a clean, reliable fuel once it is produced because the combustion of hydrogen produces water vapor. Hydrogen can react with oxygen to form water in an exothermic combustion reaction. Hydrogen combustion does not emit toxic greenhouse gases like carbon monoxide or carbon dioxide. When hydrogen is burned in air, it does produce traces of nitrogen oxides, which are pollutants. A particularly costly problem with hydrogen is that hydrogen is not readily available as free hydrogen; hydrogen is most readily available in combination with other elements, such as oxygen in water. Consequently, elemental hydrogen must be freed before it can be used as a fuel. The production of hydrogen is discussed below.

HYDROGEN PRODUCTION

Hydrogen can be produced by electrolysis. Electrolysis is the non-spontaneous splitting of a substance into its constituent parts by supplying electrical energy. In the case of water, electrolysis would decompose the water molecule into its constituent elements (hydrogen and oxygen) by the addition of electrical energy.

It is difficult to electrolyze very pure water because there are few ions present to flow as an electrical current. Electrolyzing an aqueous salt solution enhances the production of hydrogen by the electrolysis of water. An aqueous salt solution is a mixture of ions and water. The addition of a small amount of a non-reacting salt such as sodium sulfate accelerates the electrolytic process. The salt is called an electrolyte and provides ions that can flow as a current.

We illustrate the electrolysis process in Figure 7-2 using the electrolysis of water-salt solution. The electrolytic cell is a cell with two electrodes: the anode and the cathode. Oxidation occurs at the anode and reduction occurs at the cathode. The voltage source in Figure 7-2 provides the electric potential difference needed to initiate and support the electrolysis reaction. In the figure, a negatively charged chlorine ion with one extra electron loses its excess electron at the anode and combines with another chlorine atom to form the chlorine molecule (Cl_2). The electrons flow from the anode to the cathode where a positively charged sodium ion gains an electron and becomes an electrically neutral atom of sodium.

The electrolysis reaction can be treated as two half-reactions. The first half-reaction is oxidation at the anode. Oxidation occurs when electrons are released by the reducing agent in the oxidation half-reaction. A reducing agent is a substance that donates electrons in a reaction. The hydrogen atom from a water molecule is the reducing agent at the anode. The second half-reaction is reduction at the cathode. Reduction occurs when the oxidizing agent acquires electrons in the reduction half-reaction. An oxidizing agent is a substance that accepts electrons in a reaction. The hydroxyl radical OH^- from a water molecule is the oxidizing agent at the cathode for the upper reaction and the sodium ion is the oxidizing agent for the lower reaction.

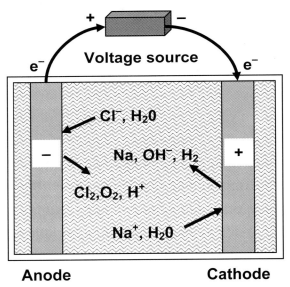

Figure 7-2. Electrolysis of Water-Salt Solution

The overall reaction for electrolysis of a water-salt solution in an electrolytic cell shows that hydrogen gas is produced at the cathode of the electrolytic cell and oxygen gas is produced at the anode when enough electrical energy is supplied. Four hydrogen ions combine with four hydroxyl ions to form four molecules of water. Electrical energy is supplied in the form of an electrical voltage and must be greater than the threshold energy (activation energy) of the reaction. The voltage needed to electrolyze water must be greater than the voltage that we would predict if we did not consider activation energy, consequently the actual voltage needed to electrolyze water is called overvoltage.

Point to Ponder: Why is electrolysis important?

For our purposes, electrolysis can produce hydrogen for use as a fuel. When electrolysis is completed, we have produced hydrogen and oxygen from water molecules. The distribution and amount of water on Earth implies that we have an abundant source of hydrogen for energy use. We can collect and use the hydrogen as a fuel. The energy needed to separate water into hydrogen and oxygen is greater than the energy provided by the produced hydrogen. That is why electrolysis is considered a net energy consuming process and hydrogen is considered an energy carrier rather than an energy source.

Thermal Decomposition and Gasification

Electrolysis is one technique for producing hydrogen. Two other techniques are thermal decomposition and gasification. The process of decomposing the water molecule at high temperatures is called thermal decomposition. Heat in the form of steam can be used to reform hydrocarbons and produce hydrogen.

Steam reforming exposes a hydrocarbon such as methane, natural gas, or gasoline to steam at high temperature and pressure [Sørensen, 2000, pages 572-573]. Steam reforming reactions produce hydrogen. One steam reforming reaction uses hydrocarbons, which we have seen have limited long-term potential because hydrocarbons are not considered a renewable source of energy. We can produce a hydrocarbon like natural gas using a technique such as methane digestion, but the rate of production is relatively slow compared to the current rate of consumption. A process such as absorption or membrane separation can be used to remove the carbon dioxide byproduct. The production of carbon dioxide, a greenhouse gas, is an undesirable characteristic of steam reforming.

Some bacteria can produce hydrogen from biomass by fermentation or high-temperature gasification. The gasification process is similar to coal gasification described in Chapter 6.

Point to Ponder: Where do we get the energy we need to produce hydrogen?

One of the problems facing society is how to produce hydrogen using an environmentally acceptable process. Three possible sources of energy for producing hydrogen in the long-term are wind energy, nuclear fusion, and solar energy. We have already seen that wind energy and solar energy are not always produced when there is a demand for energy. One way to enhance the usefulness of wind and solar energy is to use excess wind and solar energy to produce hydrogen during periods when the wind is blowing and the Sun is shining. The produced hydrogen could then be used as a source of energy when wind energy and solar energy are not being generated.

J.H. Ausubel [2000] has suggested that the potential of nuclear energy will be realized when nuclear energy can be used as a source of electricity and high-temperature heat for splitting water into its constituent parts. In this scenario, nuclear energy could be generated at sites with excess energy generating capacity and then used to produce hydrogen. The hydrogen would then fulfill its role as a carrier of energy.

HYDROGEN AND FUEL CELLS

Hydrogen is considered a carrier of energy because it can be transported in the liquid or gaseous state by pipeline or in cylinders. Once produced and distributed, hydrogen can be used as a fuel for a modified internal combustion engine or as the fuel in a fuel cell. Fuel cells are electrochemical devices that directly convert hydrogen, or hydrogen-rich fuels, into electricity using a chemical rather than a combustion process.

A fuel cell consists of an electrolyte sandwiched between an anode and a cathode (Figure 7-3). The electrolyte in Figure 7-3 is a mixture of potassium hydroxide and water. The electrolyte solution is maintained at a lower pressure than the gas cavities on either side of the porous electrodes. The pressure gradient facilitates the separation of hydrogen and oxygen molecules. The load in the figure is a circuit with amperage A and voltage V. Hydrogen is fed to the anode (negative electrode) and oxygen is fed to the cathode (positive electrode). When activated by a catalyst, hydrogen atoms separate into protons and electrons. The charged protons and electrons take different paths to the cathode. The electrons go through the external circuit and provide an electrical current, while the protons migrate through the electrolyte to the cathode. Once at the cathode, the protons combine with electrons and oxygen to produce water and heat.

The type of electrolyte in a fuel cell distinguishes that fuel cell from other types of fuel cells. The fuel cell in Figure 7-3 is a proton exchange membrane fuel cell because it depends on the movement of protons (hydrogen nuclei) through the porous electrodes. Fuel cells produce clean energy in the form of electricity and heat from hydrogen. Fuel cells do not need recharging or replacing and can produce electricity as long as they are supplied with hydrogen and oxygen.

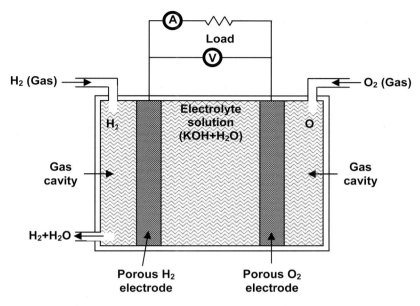

Figure 7-3. Schematic of a Fuel Cell

THE HYDROGEN ECONOMY

The historical trend toward decarbonization reflects the contention by many energy forecasters that hydrogen will be the fuel of choice in the future. These forecasters believe that power plants and motor vehicles will run on hydrogen. The economies that emerge will depend on hydrogen and are called hydrogen economies. The concept of a hydrogen economy is not new. The use of hydrogen as a significant fuel source driving a national economy was first explored in the middle of the 20th century as a complement to the adoption of large-scale nuclear electric generating capacity. Concerns about global climate change and the desire to achieve sustainable development have renewed interest in hydrogen as a fuel.

A future that depends on hydrogen is not inevitable. Hydrogen economies will require the development of improved technologies for producing, storing, transporting, and consuming hydrogen. We have already discussed some of the challenges involved in the production of hydrogen. As another example of the technological challenges that must be overcome in a transition to a hydrogen economy, let us consider the storage of hydrogen.

Hydrogen can be stored in the liquid or gaseous state, but it must be compressed to high pressures or liquefied to achieve reasonable storage volumes because of the low density of the diatomic hydrogen molecule. The energy content of hydrogen gas is less than the energy contained in methane at the same temperature and pressure. The volumetric energy density of liquid hydrogen is approximately 8700 megajoules/meter^3. This is about one third the volumetric energy density of gasoline. The relatively low volumetric energy density of hydrogen creates a storage problem if we want to store hydrogen compactly in vehicles.

Researchers have learned that hydrogen can be stored effectively in the form of solid metal hydrides. A metal hydride is a metal that absorbs hydrogen. The hydrogen is absorbed into the spaces, or interstices, between atoms in the metal. According to M. Silberberg [1996, page 246], metals such as palladium and niobium "can absorb 1000 times their volume of H_2 gas, retain it under normal conditions, and release it at high temperatures." This form of storage may be desirable for use in hydrogen-powered vehicles. A model of a car using hydrogen fuel cells is shown in Figure 7-4.

Figure 7-4. Hydrogen Fuel Cell Car, Denver, Colorado Exhibit

Hydrogen can be hazardous to handle. A spectacular demonstration of this fact was the destruction of the German zeppelin *Hindenburg*. The *Hindenburg* used hydrogen for buoyancy. In 1937, the *Hindenburg* burst into flames while attempting a mooring in Lakehurst, New Jersey. At the time, people believed that the hydrogen in the *Hindenburg* was the cause of the explosion. Addison Bain, at the end of the 20th century, showed that the chemical coating on the outside of the zeppelin was the cause of the explosion. When the chemical coating ignited, the hydrogen began to burn. Today, lighter-than-air ships use less flammable gases such as helium.

Hydrogen forms an explosive mixture with air when the concentration of hydrogen in air is in the range of 4% to 75% hydrogen. For comparison, natural gas is flammable in air when the concentration of natural gas in air is in the range of 5% to 15% natural gas. Furthermore, the energy to ignite hydrogen-air mixtures is approximately one-fifteenth the ignition energy for natural gas-air or gasoline-air mixtures. The flammability of hydrogen in air makes it possible to consider hydrogen a more dangerous fuel than natural gas. On the other hand, the low density of hydrogen allows hydrogen to dissipate more quickly into the atmosphere than a higher density gas such as methane. Thus, hydrogen leaks can dissipate more rapidly than natural gas leaks. Adding an odorant to the gas can enhance the detection of gas leaks.

The environmental acceptability of hydrogen fuel cells depends on how the hydrogen is produced. If a renewable energy source such as solar energy is used to generate the electricity needed for electrolysis, vehicles powered by hydrogen fuel cells would be relatively clean since hydrogen combustion emits water vapor. Unfortunately, hydrogen combustion in air also emits traces of nitrous oxide compounds. Nitrogen dioxide contributes to photochemical smog and can increase the severity of respiratory illnesses.

Point to Ponder: Would people be willing to use hydrogen in their everyday lives?

The common conception of hydrogen depends on a person's cultural background. People in the United States tend to think of the *Hindenburg* on fire when they think of hydrogen as a fuel. One of the tasks of an evolving hydrogen industry is to educate the public about the misconceptions of hydrogen.

One example of the education effort is to re-educate the public about the *Hindenburg* disaster. Hydrogen economy proponents have presented Bain's new explanation of the *Hindenburg* disaster at trade shows and public exhibitions. Videos of staged automobile accidents are another example of the education effort.

The videos show how a hydrogen-fueled car would burn following an accident compared to a gasoline-fueled car. Hydrogen burns like a flare pointing to the sky because hydrogen is less dense than air. By contrast, gasoline from a leaking fuel tank pools on the ground beneath the car and can engulf the car in flames. The videos show that hydrogen could be used as a relatively safe fuel in transportation. These examples show that proponents of the hydrogen economy believe that an education effort is needed to overcome social resistance to adopting hydrogen as a primary fuel.

Chapter 8

ELECTRICITY GENERATION
AND DISTRIBUTION

Power plants electrified the United States, and eventually the rest of the modern world, in only a century and a half. Although other types of energy are used around the world, electricity is the most versatile form for widespread distribution. The role of electric power plants is to generate electric current for distribution through a transmission grid. The historical developments that led to the modern power generation and distribution system are described below. We then discuss some changes that are being implemented to prepare electric distribution systems for the emerging energy mix.

HISTORICAL DEVELOPMENT OF ELECTRIC POWER

People first used muscle energy to gather food and build shelters. Muscle energy was used to grind grain with stones, chop wood with hand axes, and propel oar-powered ships. In many instances in history, conquered people became slaves and provided muscle energy for their conquerors.

Stones, axes, and oars are examples of tools that were developed to make muscle energy more effective. Water wheels and windmills replaced muscle power for grinding grain as long ago as 100 B.C.E. Wind and sails replaced muscle energy and oar-powered ships. Early power stations were driven by wind and flowing water, and were built where wind and flowing water were available. Steam generated power allowed power plants to move, but had to await the development of furnaces.

Steam Generated Power

Furnaces used heat to smelt ore. Ore is rock that contains metals such as copper, tin, iron, and uranium. Heat from fire frees the metal atoms and allows them to be collected in a purified state. Copper and tin were the first metals smelted and could be combined to form bronze.

Heat and water were combined to generate steam and steam engines were developed to convert thermal energy to mechanical energy. Early steam engines drove a piston that was placed between condensing steam and air, as illustrated in Figure 8-1. When steam condenses, it occupies less volume and creates a partial vacuum. The air on the other side of the piston expands and can push the piston. By alternately injecting steam and letting it condense, the piston can be made to move in an oscillating linear motion.

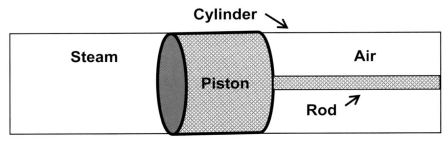

Figure 8-1. Schematic of a Simple Steam Engine

English inventor Thomas Newcomen invented the steam engine in 1705 and built the first practical steam engine in 1712. Newcomen's steam engine was used to pump water from flooded coal mines. Steam condensation was induced in Newcomen's steam engine by spraying cold water into the chamber containing steam. The resulting condensation creates the partial vacuum that allows air to push the piston. A weight attached to the rod used gravity to pull the piston back as steam once again entered the left hand chamber in Figure 8-1.

Scottish engineer James Watt improved the efficiency of the steam engine by introducing the use of a separate vessel for collecting and condensing the expelled steam. Watt's assistant, William Murdock, developed a gear system design in 1781 that enabled the steam engine to produce circular motion. The ability to produce circular motion made it possible for steam engines to provide the power needed to turn wheels. Steam engines could be placed on platforms

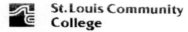

St. Louis Community College

Current Check-Outs summary for ZOU, MEI
Wed Apr 06 11:52:59 CDT 2011

BARCODE: 300080005980769
TITLE: Energy in the 21st century / John
DUE DATE: Apr 27 2011
STATUS:

BARCODE: 300080006475670
TITLE: Renewable energy : sources and me
DUE DATE: Apr 27 2011
STATUS:

attached to wheels to provide power for transportation. Thus was born the technology needed to develop steam-driven locomotives, paddle wheel boats, and ships with steam-driven propellers. Furthermore, steam engines did not have to be built near a particular fuel source. It was no longer necessary to build manufacturing facilities near sources of wind or water, which were formerly used to provide power. Manufacturers had the freedom to build their manufacturing facilities in locations that optimized the success of their enterprise. If they chose, they could build near coal mines to minimize fuel costs, or near markets to minimize the cost of distributing their products.

Steam generated power was an environmentally dirty source of power. Burning biomass such as wood or a fossil fuel such as coal typically produced the heat needed to generate steam. Biomass and fossil fuels were also used in the home. Attempts to meet energy demand by burning primarily wood in 16th century Britain led to deforestation and the search for a new fuel source [Nef, 1977]. Fossil fuel in the form of coal became the fuel of choice in Britain and other industrialized nations. Coal gas, which is primarily methane, was burned in 19th century homes.

The demand for energy had grown considerably by the 19th century. Energy for cooking food and heating and lighting homes was provided by burning wood, oil, or candles. The oil was obtained from such sources as surface seepages or whale blubber. Steam generated power plants could only serve consumers in the immediate vicinity of the power plant. A source of power was needed that could be transmitted to distant consumers.

By 1882, Thomas Edison was operating a power plant in New York City. Edison's plant generated direct current electricity at a voltage of 110 volts. Nations around the world soon adopted the use of electricity. By 1889, a megawatt electric power station was operating in London, England. Industry began to switch from generating its own power to buying power from a power generating company. But a fundamental inefficiency was present in Edison's approach to electric power generation. The inefficiency was not re-moved until the Battle of Currents was fought and won.

The Battle of Currents

The origin of power generation and distribution is a story of the Battle of Currents, a battle between two titans of busi-ness: Thomas Edison and George Westinghouse. The motivation for their confrontation can be reduced to a single, fundamental issue: how to electrify America.

Edison invented the first practical incandescent lamp and was a proponent of electrical power distribution by direct electric current. He displayed his direct current technology at New York City's Pearl Street Station in 1882. One major problem with direct current is that it cannot be transmitted very far without significant degradation.

Unlike Edison, Westinghouse was a proponent of al-ternating electric current because it could be transmitted over much greater distances than direct electric current.

Alternating current could be generated at low voltages, transformed to high voltages for transmission through power lines, and then reduced to lower voltages for delivery to the consumer. Nikolai Tesla (1857-1943), a Serbian-American scientist and inventor who was known for his work with magnetism, worked with Westinghouse to develop alternating current technology. Westinghouse displayed his technology at the 1893 Chicago World's Fair. It was the first time one of the world's great events was illuminated at night and it showcased the potential of alternating current electricity.

The first large-scale power plant was built at Niagara Falls near Buffalo, New York in the 1890's. The power plant at Niagara Falls began transmitting power to Buffalo, less than 30 kilometers (20 miles) away, in 1896. The transmission technology used alternating current technology. The superiority of alternating current technology gave Westinghouse a victory in the battle of currents and Westinghouse became the father of the modern power industry. Westinghouse's success was not based on better business acumen, but on the selection of better technology. The physical principles that led to the adoption of alternating current technology are discussed below.

A chronology of milestones in the development of electrical power is presented in Table 8-1 [after Brennan, et al., 1996, page 22; and Aubrecht, 1995, Chapter 6]. The milestones refer to the United States, which was the worldwide leader in the development of an electric power industry.

Table 8-1. Early Milestones in the History of the Electric Power Industry in the United States		
Year	Event	Comment
1882	Pearl Street Station, New York	Edison launches the "age of electricity" with his DC power station
1893	Chicago World's Fair	Westinghouse displays AC power to the world
1898	Fledgling electric power industry seeks monopoly rights as regulated utilities	Chicago Edison's Samuel Insull leads industry to choose regulation over "debilitating competition"
1907	States begin to regulate utilities	Wisconsin and New York are first to pass legislation
1920	Federal government begins to regulate utilities	Federal Power Commission formed

Growth of the Electric Power Industry

The power industry started out as a set of independently owned power companies. Because of the large amounts of money needed to build an efficient and comprehensive electric power infrastructure, the growth of the power industry required the consolidation of the smaller power companies into a set of fewer but larger power companies. The larger, regulated power generating companies became public utilities and could afford to build regional electric power transmission grids. The ability to function more effectively at larger scales is an example of an economy of scale.

Utility companies were able to generate and distribute more power at lower cost by building larger power plants and transmission grids.

The load on a utility is the demand for electrical power. Utilities need to have power plants that can meet three types of loads: base load, intermediate or cycling load, and peak load. The base load is the minimum baseline demand that must be met in a 24-hour period. Intermediate load is the demand that is required for several hours each day and tends to increase or decrease slowly. Peak load is the maximum demand that must be met in a 24-hour period.

Electric power for small towns and rural communities was an expensive extension of the power transmission grid that required special support. The federal government of the United States provided this support when it created the Tennessee Valley Authority (TVA) and Rural Electric Associations (REA).

Point to Ponder: Is society willing to adopt new technologies?

The history of electric power development shows that society is willing to adopt new technologies. We have seen this recently in the adoption of the world wide web, or internet, and cell phones. When people understand the benefits of new technology and an appropriate infrastructure is created, people will adopt the technology.

ELECTRIC POWER GENERATION

The first commercial-scale electric power plants were hydroelectric plants. The primary energy source, or the energy that is used to operate an electricity-generating power plant, is flowing water for a hydroelectric plant. Today, most electricity is generated by one of the following primary energy sources: coal, natural gas, oil, or nuclear. Table 8-2 presents the consumption of primary energy in the year 1999 as a percentage of total primary energy consumption in the world for a selection of primary energy types.

Table 8-2. Primary Energy Consumption in 1999 by Energy Type [Source: EIA website, 2002]	
Primary Energy Type	**Total World Energy Consumption**
Oil	39.9 %
Natural Gas	22.8 %
Coal	22.2 %
Hydroelectric	7.2 %
Nuclear	6.6 %
Geothermal, Solar, Wind and Wood	0.7 %

Table 8-2 is based on statistics maintained at a website by the Energy Information Administration (EIA), United States Department of Energy. The statistics should be considered approximate. They are quoted here because the EIA is a standard source of energy information that is widely referenced. The statistics give us an idea of the relative importance of different primary energy sources. Fossil fuels were clearly the dominant primary energy source at the end of the 20th century. Electric energy, however, is the most versatile source of energy for running the 21st century world and much of the primary energy is consumed in the generation of electric energy.

An example of electric power generation is hydroelectric power generation. Many of the first commercial electric power plants relied on flowing water as their primary energy source. People have known for some time that falling water could be used to generate electric power. A schematic of a hydroelectric power plant is presented in Figure 5-4.

Figure 8-2. Hoover Dam Generators, Nevada

Water flows from an upper elevation to a lower elevation through a pipeline called a penstock. The water current turns a turbine that is connected to a generator. Figure 8-2 shows a row of electric power generators at Hoover Dam, Nevada. The mechanical energy of falling water is transformed into the kinetic energy of rotation of the turbine. The kinetic energy of the rotating turbine is converted to electrical energy by an alternating current generator. The flowing water that turned the turbine is expelled on the downstream side of Hoover Dam (Figure 8-3).

Figure 8-3. Downstream of the Hoover Dam, Nevada

An alternating current generator converts mechanical energy to electrical energy using the following principle: an electric current is induced in a loop of wire that is allowed to rotate inside a constant magnetic field. If falling water is allowed to strike a turbine that is attached to a coil of wire inside a magnet, the coil of wire will rotate inside the magnetic field and an electric current will be induced in the coil of wire. An electromotive force, or electric potential, is associated with the electric current. The electromotive force can be used to transmit electrical power through transmission lines. To do this efficiently, a transformer must be included in the system.

TRANSFORMERS

Transformers had to be built that could be used with alternating current. A transformer is a device that can convert a small alternating current voltage to a larger alternating current voltage, or vice versa. For example, it is desirable to work with relatively small voltages at the power plant and provide a range of voltages to the consumer. In between, in the transmission lines, a large voltage is required to minimize resistive heating losses in the line. Transformers perform the function of converting, or transforming, voltages.

A transformer that increases the voltage is referred to as a step-up transformer and is said to step-up the voltage; a transformer that decreases the voltage is referred to as a step-down transformer and is said to step-down the voltage. Transformer T_1 in Figure 8-4 is a step-up transformer from the low voltage L.V. at the power station to a higher voltage H.V. in the transmission line. Transformer T_2 in the figure is a step-down transformer that converts the relatively high voltage in the transmission line to a lower voltage that is suitable for the consumer. The low voltages shown at opposite ends of the transmission line do not have to be the same. The actual voltages used in the transmission line depend on the properties of the transformer. Typical transmission line voltages can range from under 100,000 volts to over 750,000 volts [Wiser, 2000, page 197].

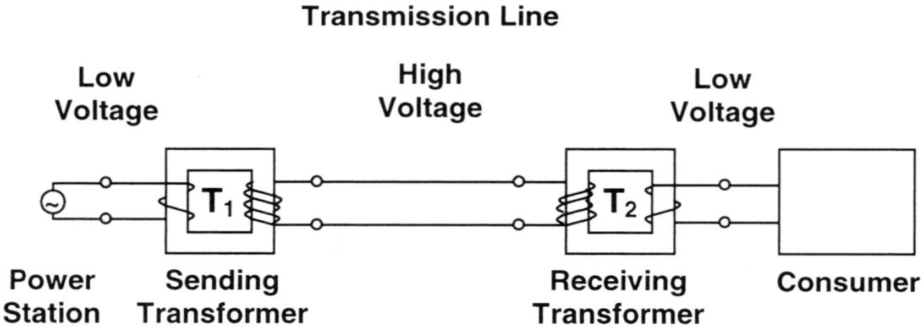

Figure 8-4. Power Transmission

Point to Ponder: What is the advantage of reducing the current in an electrical transmission line?

Power loss in an electrical transmission line depends on current. A reduction in current will reduce power loss and increase the efficiency of distributing electrical power to the consumer.

ELECTRIC POWER DISTRIBUTION

Transmission lines are used to distribute electric power. As a rule, society would like to minimize power loss due to heating to maximize the amount of primary energy reaching the consumer from power plants. We can reduce power loss by reducing the current or by decreasing the distance of transmission. In most cases, it is not a viable option to decrease the transmission distance. It is possible, however. For example, you could choose to build a manufacturing facility near a power station to minimize the cost of transmission of power. One consequence of that decision is that the manufacturer may incur an increase in the cost of transporting goods to market.

A more viable option for reducing power loss is to reduce the current that must be transmitted through transmission lines. Power loss increases faster with an increase in current than with an increase in transmission distance. This explains why Edison's direct current concept was not as attractive as Westinghouse's alternating current concept. Transformers do not work with direct current since there is no time-varying magnetic flux, so the transmission of direct current incurs resistive power losses based on the direct current generated at the power plant. The purpose of transformers is to reduce power losses in the transmission line by reducing the current in the transmission line. This is possible with alternating current.

Point to Ponder: How far can electrical power be transmitted?

Electrical power can be transmitted over great distances through transmission lines. For example, the eastern half of the United States uses an interconnected electricity distribution grid. A failure in power transmission in this grid can affect large geographic areas and many people. For example, separate power transmission failures in 2003 along the United States – Canadian border and throughout Italy resulted in the loss of electrical power for millions of people. Power losses in transmission lines limit the distance that electrical power can be transmitted. An option for the future is to use superconductors as transmission lines.

Superconductors are materials that offer no resistance to electron flow. Dutch physicist Hans Kamerlingh Onnes (1853-1926) discovered the first superconductor in 1911. Onnes produced liquid helium by cooling helium to a temperature below 4.2° Kelvin. He then observed that the resistivity of mercury vanished when mercury was cooled by liquid helium. Superconducting materials have been developed that operate at higher temperatures than 4.2° Kelvin, but superconductors can still only operate at temperatures that are far below room temperature. Super-conductors are not yet feasible for widespread use in power transmission because they require costly refrigeration. [Fanchi, 2004, Exercise 2-8]

Transmission and Distribution System

Electric power generating stations are connected to loads (consumers) by means of a transmission system that consists of transmission lines and substations. The substations are nodes in the transmission grid that route electric power to loads at appropriate voltages. Figure 8-5 shows the basic elements of an electric power transmission system. It includes transmission lines, transformers, and distribution systems.

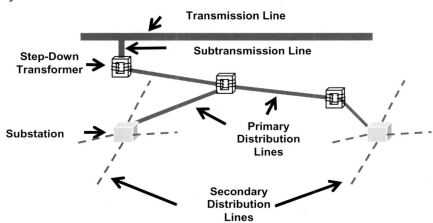

Figure 8-5. Power Transmission System

Typical transmission voltages in the United States range from 69 kilovolts to 765 kilovolts, and the alternating current frequency is 60 hertz. The infrastructure for providing electric power to the loads from the substations is the distribution system. A failure of the power transmission system can leave millions of people without power, as it did in the 2003 blackouts in the northeastern part of North America and Italy.

Electric power is transmitted and distributed cost-effectively by operating a three-phase system. Three-phase electricity refers to current and voltages that are out of phase with each other by 120 degrees. A three-phase alternating current generator is used to provide three-phase electricity. Transmission lines for three-phase electricity are shown in Figure 8-6A, and a typical pole-mounted transformer is shown in Figure 8-6B.

**Figure 8-6A. Transmission of Three-Phase Electricity;
8-6B. Pole-Mounted Transformer**

Three-phase electricity can be distributed to three one-phase loads using three separate conductors. Three-phase transmission lines are designed to operate at high voltages and low currents. High voltage transformers are used to step-down the voltage for use by consumers.

The pole-mounted transformer shown in Figure 8-6B can convert three-phase power to single-phase voltages that are suitable for consumers. Residential consumers in the United States typically use single-phase voltages of 120 volts for small appliances and 240 volts for larger appliances such as clothes dryers. Large, industrial consumers can use three-phase voltages on the order of 2160 volts or higher.

Household Circuits

Households are among the most common consumers of electricity in the modern world. Electricity is delivered as alternating current to a typical house in the United States using either a two-wire line, or a three-wire line. The potential difference, or root mean square voltage, between the two wires in the two-wire line in the United States is 120 volts, and is 240 volts in many parts of Europe. One of the wires in the two-wire line is connected to a ground at the transformer and the other wire is the "live" wire. The three-wire line has a neutral line, a line at +120 volts, and a line at -120 volts.

A meter is connected in series to the power lines to measure the amount of electricity consumed by the household. The meter provides the information the utility needs to bill the consumer. In addition to the meter, a circuit breaker is connected in series to the power lines to provide a safety buffer between the house and the power line. A fuse may be used instead of a circuit breaker in older homes. The fuse contains a metal alloy link, such as lead-tin, with a low melting temperature. If the alloy gets too hot because of resistive heating, the link will melt and break the circuit. More modern circuit breakers are electromagnetic devices that use a bi-metallic strip. If the strip gets too hot, the bimetallic strip will curl based on the coefficients of thermal expansion of the two metals that make up the bimetallic strip. The curling strip will break the circuit.

The power line and circuit breaker are designed to handle the current in the circuit, which can be as high as 30 amps, although many household applications require 15 amps of current. Lamps and appliances such as microwaves and toasters operate at 120 volts, while electric ranges and clothes dryers operate at 240 volts. Each circuit in the house has a circuit breaker or fuse to accommodate different loads.

The circuit in the house is connected in parallel to the power lines. The parallel connection makes it possible to turn on and shut off an electrical device without interfering with the operation of other electrical devices. If a series connection was used, the circuit would be broken whenever one of the electrical devices was turned off. That is why circuit breakers are connected in series between the household circuit and the power lines; the circuit breaker is designed to disconnect electrical devices in the house from the power lines in the event of an overload, such as a power surge. An open circuit occurs when the continuity of the circuit is broken. A short circuit occurs when a low resistance pathway is created for the current.

Electricity can be harmful if a person touches a live wire while in contact with a ground. Electric shocks can cause burns that may be fatal and can disrupt the routine functioning of vital organs such as the heart. The extent of biological damage depends on the duration and magnitude of current. A current in excess of 100 milliamps can be fatal if it passes through a body for a few seconds.

Electrical devices and power lines should be handled with care. Three-pronged power cords for 120-volt outlets provide two prongs that are grounded and one prong that is connected to the live wire. The grounded prongs are provided for additional safety in electrical devices designed to use the three-pronged cords. One of the ground wires is connected to the casing of the appliance and provides a low resistance pathway for the current if the live wire is short-circuited.

Distributed Generation

Practical considerations limit the size of power plants. Most large-scale power plants have a maximum capacity of approximately 1000 megawatts. The size of the power plant is limited by the size of its components, by environmental concerns, and by energy source. For example, the area occupied by a power plant, called the plant footprint, can have an impact on land use. Conventional power plants that burn fossil fuels such as coal or natural gas can produce on the order of 1000 megawatts of power. Power plants that depend on nuclear reactors also produce on the order of 1000 megawatts of power. By contrast, power plants that rely on solar energy presently can produce on the order of 10 megawatts of power. Power from collections of wind turbines in wind farms can vary from one megawatt to hundreds of megawatts. If we continue to rely on nuclear or fossil fuels, we need to work with power plants that generate on the order of 1000 megawatts of power. If we switch to power plants that depend on solar energy or wind energy, the power generating capacity of each plant is less than 1000 megawatts of power and we must generate and transmit power from more plants to provide existing and future power needs.

In some areas, public pressure is growing to have more power plants with less power generating capacity and more widespread distribution. The federal government of the United States passed a law in 1978 called the Public Utilities Regulatory Policies Act (PURPA) that allows non-utilities to generate up to 80 megawatts of power and requires utilities to purchase this power. PURPA was the first law passed in decades to relax the monopoly on power generation held by utilities and reintroduce competition in the power generating sector of the United States economy.

The distributed generation of energy is the generation of energy where it is needed and on a scale that is suitable for the consumer. Examples of distributed generation include a campfire, a wood stove, a candle, a battery-powered watch, and a car. Each of these examples generates its own power for its specific application. Historically, distributed generation was the first power generation technology. The electric power generation and transmission grid that emerged in the 20th century and is still in use today is a centralized system that relies on large-scale power generating plants and extensive transmission capability. The transmission grid provides power to distant locations.

Some people believe that the future of energy depends on a renaissance in distributed generation. In this view, a few large-scale power plants in the centralized system will be replaced by many smaller-scale power-generating technologies. A. Borbely and J.F. Kreider define distributed generation as "power generation technologies below 1 megawatt of electrical output that can be sited at or near the load they serve" [Borbely and Kreider, 2001, page 2]. This definition does not include small-scale power-generating technologies whose ideal locations depend on the locations of their energy source. For example, hydropower and wind-powered generators are not considered distributed generation technologies according to Borbely and Kreider's definition because hydropower and wind-powered generators depend on the availability of flowing water and air respectively. Consequently, hydropower and wind-powered generators must be located near their energy sources and these locations are often not near the power consumer.

Point to Ponder: Can distributed generation increase the usefulness of wind and solar energy?

We have noted previously that energy from wind and sunshine is not always produced when it is needed. The Sun does not shine on a particular spot on the surface of the Earth all day long, but the Sun does shine all the time.

We can use sunlight 24 hours per day if we can collect sunlight at different locations around the globe. Similarly, the wind does not usually blow all the time at a particular location, but wind is always blowing somewhere. Distributed generation can be used to harvest both wind energy and solar energy. Improved power distribution and transmission systems could be used to produce energy from wind and sunshine in different parts of the world and transmit it to places where it is needed. A global energy distribution system would have to be developed and maintained to achieve this capability.

Chapter 9

ENERGY, ECONOMICS, AND THE ENVIRONMENT

Energy may be the most important factor that will influence the shape of society in the 21st century. The cost and availability of energy significantly impacts our quality of life, the health of national economies, the relationships between nations, and the stability of our environment. What kind of energy do we want to use in our future? Will there be enough? What will be the consequences of our decisions? The selection of an energy source depends on such factors as availability, accessibility, environmental acceptability, capital cost, and ongoing operating expenses. We discuss these factors here.

ENERGY AND THE ECONOMY

One essential component of energy resource management is economics. An economic analysis is used to weigh various options and decide on an appropriate course of action. We provide a brief introduction to economics in this section.

Cash Flow and Economic Indicators

An economic analysis of competing investment options usually requires preparing cash flow predictions. The cash flow of an investment option is the net cash generated by the investment option or expended on the investment option as a function of time (see Figure 9-1).

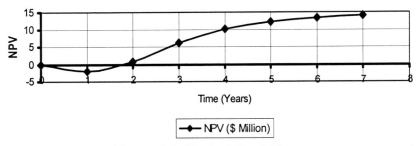

Figure 9-1: Typical Cash Flow

Net present value on the vertical axis of Figure 9-1 is the difference between revenue and expenses. Net present value, revenue, and expenses depend on the time value of money. We can account for the time value of money by introducing a discount rate in the calculation. Revenue can be calculated from the price per unit quantity produced times the quantity produced. The quantity produced can be volume of oil or gas, kilowatt-hours of electricity, or any other appropriate measure of resource production. Expenses include capital expenditures, operating expenditures, and taxes.

The time dependence of NPV is illustrated in Figure 9-1. This figure shows that NPV can be negative. A negative NPV says that the investment option is operating at a loss. The loss is usually associated with initial capital investments and operating expenses that are incurred before the invest-ment option begins to generate revenue. For example, mirrors have to be installed in a solar electric generating system and a generator built before electricity can be sold to the transmission grid. The installation of mirrors and a generator are capital expenses that occur before revenue is generated by the sale of electricity. Similarly, large invest-ments in the design and construction of offshore platforms must be made before wells can be drilled and oil produced. The eventual growth in positive NPV occurs when revenue exceeds expenses. The point in time when NPV equals zero is the payout time. In the example shown in Figure 9-1, payout time is approximately 1.5 years. The concept of payout time can be applied to discounted cash flow or undiscounted cash flow.

The time value of money is included in the economic analyses by applying a discount rate to adjust the value of money in future years to the value of money during a base year. The resulting cash flow is called the discounted cash flow. The net present value (NPV) of the discounted cash flow is the value of the cash flow at a specified discount rate.

Several commonly used indicators of economic per-formance are listed in Table 9-1. The discount rate at which the maximum value of NPV is zero is called the discounted cash flow return on investment (DCFROI) or Internal Rate of Return (IRR). Payout time, NPV, and DCFROI account for the time value of money. An indicator that does not account for the time value of money is the profit-to-investment (PI) ratio. The PI is the total undiscounted cash flow without capital investment divided by total investment. It is often useful to prepare a variety of plots, such as NPV versus time and NPV versus discount rate, to show how economic indicators perform as functions of time.

Table 9-1. Indicators of Economic Performance	
Discount Rate	Factor to adjust the value of money to a base year.
Net Present Value	Value of cash flow at a specified discount rate.
DCFROI or IRR	Discount rate at which net present value is zero.
Payout Time	Time when net present value equals zero.
Profit-to-Investment Ratio	Undiscounted cash flow without capital investment divided by total investment.

Economic analyses using indicators of economic performance provide information about the relative perform-ance of different investment options. The economic viability of an investment option is usually decided after considering a combination of economic indicators. For example, an investment option may be considered economically viable if the DCFROI is greater than 30% and payout time is less than two years. It should be remembered, however, that quantitative indicators provide useful information, but not complete information.

Economic viability is influenced by both tangible and intangible factors. The tangible factors, such as building a generator or drilling a well, are relatively easy to quantify. Intangible factors such as environmental and socio-political concerns are relatively difficult to quantify, yet may be more important than tangible factors.

Life Cycle Analysis

The analysis of the costs associated with an energy source should take into account the initial capital expenditures and annual operating expenses for the life of the system. This analysis is called life cycle analysis and the costs are called life cycle costs. Life cycle costing requires the analysis of all direct and indirect costs associated with the system for the entire expected life of the system. According to Bent Sørensen [2000, page 762], life cycle analysis includes analyzing the impact of "materials or facilities used to manufacture tools and equipment for the process under study and it includes final disposal of equipment and materials, whether involving reuse, recycling or waste disposal." A list of life cycle costs is presented in Table 9-2.

Table 9-2: Life Cycle Costs of an Energy System [Goswami, et al., 2000, page 528]
Capital equipment costs
Acquisition costs
Operating costs for fuels, etc.
Interest charges for borrowed capital
Maintenance, insurance, and miscellaneous charges
Taxes (local, state, federal)
Other recurring or one-time costs associated with the system
Salvage value (usually a cost) or abandonment cost

It is important to recognize that the future cost of some energy investment options may change significantly as a result of technological advances. The cost of a finite resource can be expected to increase as the availability of the resource declines, while the cost of an emerging technology will usually decline as the infrastructure for supporting the technology matures.

The initial costs of one energy system may be relatively low compared to competing systems. If we only consider initial cost in our analysis, we may adopt an energy option that is not optimum. For example, the annual operating expenses for an option we might choose based on initial cost may be significantly larger than those of an alternative option. In addition, projections of future cost may be substantially in error if the cost of one or more of the components contributing to an energy system changes significantly in relation to our original estimate. To avoid making less-than-optimum decisions, we should consider all of the life cycle costs of each investment option. We also need to evaluate the sensitivity of cash flow predictions to plausible changes in cost as a function of time. Inherent in life cycle analysis is an accurate determination of end use efficiency.

End use efficiency is the overall efficiency of converting primary energy to a useful form of energy. It should include an analysis of all factors that affect the application. As a simple example, consider the replacement of a light bulb. The simplest decision is to choose the least expensive light bulb. On a more sophisticated level, we need to recognize that the purpose of the light bulb is to provide light and some light bulbs can provide light longer than other light bulbs. In this case we need to consider the life of the light bulb in addition to its price. But there are still more factors to consider. If you live in an equatorial region, you might prefer a light bulb that emits light and relatively little heat, so you can reduce air conditioning expenses. On the other hand, if you live in a cooler northern climate, you might desire the extra heat and choose a light bulb that can also serve as a heat source. Once you select a light bulb, you want to use it where it will do the most good. Thus, if you chose a more expensive light bulb that has a long life and generates little heat, you would probably prefer to use the light bulb in a room where the bulb would be used frequently, such as a kitchen or office, rather than a closet where the bulb would be used less frequently. All of these factors should be taken into account in determining the end use efficiency associated with the decision to select a light bulb.

One of the goals of life cycle analysis is to make sure that decision makers in industry, government, and society in general, are aware of all of the costs associated with a system. In the context of energy resource management, Bent Sørensen [2000, Section 7.4.3] has identified the following impact areas: economic, environmental, social, security, resilience, development, and political. Some typical questions that must be answered include the following:

1.   Does use of the resource have a positive social impact; that is, does resource use provide a product or service without adversely affecting health or work environment?

2.   Is the resource secure, or safe, from misuse or terrorist attack?

3.   Is the resource resilient; that is, is the resource relatively insensitive to system failure, management errors, or future changes in the way society assesses its impact?

4.   Does the resource have a positive or negative impact on the development of a society; that is, does the resource facilitate the goals of a society, such as decentralization of energy generating facilities or satisfying basic human needs?

5.   What are the political ramifications associated with the adoption of an energy resource?

6.   Is the resource vulnerable to political instability or can the resource be used for political leverage?

A thorough life cycle analysis will provide answers to all of these questions. Of course, the validity of the answers will depend on our ability to accurately predict the future.

Risk Analysis and Real Options Analysis

A characteristic of natural resource management is the need to understand the role of uncertainty in decision making. The information we have about a natural resource is usually incomplete. What information we do have may contain errors. Despite the limitations in our knowledge, we must often make important decisions to advance a project. These decisions should be made with the recognition that risk, or uncertainty, is present and can influence investment decisions. Here, risk refers to the possibility that an unexpected event can adversely affect the value of an asset. Uncertainty is not the same as risk. Uncertainty is the concept that our limited knowledge and understanding of the future does not allow us to predict the consequences of our decisions with 100% accuracy. Risk analysis is an attempt to quantify the risks associated with investing under uncertainty.

One of the drawbacks of traditional risk analysis is the limited number of options that are considered. The focus in risk analysis is decision making based on current expectations about future events. For example, the net present value analysis discussed previously requires forecasts of revenue and expenses based on today's expectations. Technological advances or political instabilities are examples of events that may significantly alter our expectations. We might overlook or ignore options that would have benefited from the unforeseen events. An option in this context is a set of policies or strategies for making current and future decisions. Real Options Analysis attempts to incorporate flexibility in the management of investment options that are subject to considerable future uncertainty.

The best way to incorporate options in the decision making process is to identify them during the early stages of analysis. Once a set of options has been identified for a particular project, we can begin to describe the uncertainties and decisions associated with the project. By identifying and considering an array of options, we obtain a more complete picture of what may happen as a consequence of the decisions we make. Real Options Analysis helps us understand how important components of a project, particularly components with an element of uncertainty, influence the value of the project.

Point to Ponder: Is the military budget a hidden cost of importing oil and gas?

Most industrialized nations rely on oil and gas to support their energy needs. Some of these countries, such as the United States, Great Britain, Russia and China, maintain significant military capabilities for national defense and global influence. In the case of Great Britain and the United

States, part of this military expenditure is used to maintain access to oil and gas resources. These costs should be considered part of the cost of continued reliance on fossil fuels. For example, if US$20 billion is spent to keep open supply lines to a region in one year, and you import 500 million barrels of oil from that region in a year, that would add US$40.00 to the cost of each barrel of imported oil, or about US$1.00 to each gallon of gasoline. This cost assumes that no lives are lost and, if they are, that they have no monetary value. Each society must determine the value they want to place on the lives of people who are being asked to protect their sources of energy (Figure 9-2).

Figure 9-2. Military Cemetery in Washington, D.C.

FOSSIL ENERGY AND THE ENVIRONMENT

At present, fossil fuels are the primary fuel for the majority of power plants in the world. The fuel used to drive a power plant is called the primary fuel. Examples of primary fuels include oil, coal, natural gas, and uranium. Fossil fuel resources are finite, however, and power plants that burn fossil fuels emit greenhouse gases and other pollutants. The extraction of some types of fossil fuels, such as coal and tar sands, can leave the land scarred and in need of reclamation (Figure 9-3). Before judging fossil fuels too harshly, however, it must be realized that every energy source has advantages and disadvantages.

Strip Mining

Figure 9-3. Scarred Land near Banff, Canada

Energy density, cost, and reliability are among the advantages that fossil fuels enjoy relative to other energy sources. A reliable energy source is a resource that is available almost all the time. Some downtime may be necessary for facility maintenance or other operating reasons. The

United States Department of Energy has defined a reliable energy source as an energy source that is available at least 95% of the time [DoE Geothermal, 2002].

The energy densities shown in Table 9-3 for some common materials are among the most important factors considered in selecting a fuel source. In addition to energy density, such factors as cost, reliability, social acceptability, and environmental impact must be considered.

Table 9-3. Energy Density of Common Materials*		
Material	**Energy Density**	
	MJ kg^{-1}	**MJ m^{-3}**
Crude oil	42	37,000
Coal	32	42,000
Dry wood	12.5	10,000
Hydrogen, gas	120	10
Hydrogen, liquid	120	8,700
Methanol	21	17,000
Ethanol	28	22,000
*Source: Sørensen, 2000, page 552		

Global Warming

Solar radiation heats the Earth. The average temperature at the Earth's surface is approximately 287° Kelvin (57° Fahrenheit), and typically varies from 220° Kelvin (-64° Fahrenheit) to 320° Kelvin (116° Fahrenheit) [Sørenson, 2000, page 26]. Measurements of ambient air temperature show a global warming effect that corresponds to an increase in the average temperature of the Earth's atmosphere. The increase in atmospheric temperature can be traced to the beginning of the 20th century [Lide, 2002, page 14-32] and is associated with the combustion of fossil fuels.

When a carbon-based fuel burns, carbon can react with oxygen and nitrogen in the atmosphere to produce carbon dioxide, carbon monoxide, and nitrogen oxides (often abbreviated as NOx). The combustion byproducts, including water vapor, are emitted into the atmosphere in gaseous form. Some of the gaseous byproducts are called greenhouse gases because they capture the energy in sunlight that is reflected by the Earth's surface and reradiate the energy in the form of infrared radiation. Greenhouse gases include carbon dioxide, methane, and nitrous oxide, as well as other gases such as volatile organic compounds and hydrofluorocarbons.

Global warming due to the absorption of reflected sunlight and subsequent emission of infrared radiation is called the greenhouse effect because greenhouse walls allow sunlight to enter the greenhouse and then trap reradiated infrared radiation. The greenhouse effect is illustrated in Figure 9-4. Some of the incident solar radiation from the Sun is absorbed by the Earth, some is reflected into space, and some is captured by chemicals in the atmosphere and re-radiated as infrared radiation (heat). The re-radiated energy would have escaped the Earth as reflected sunlight if green-house gases were not present in the atmosphere.

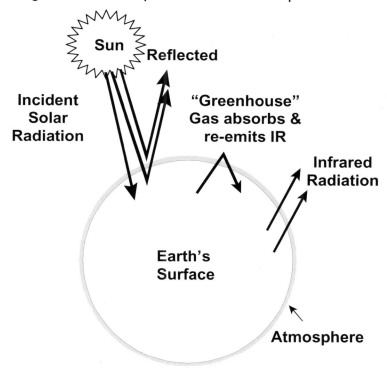

Figure 9-4. The Greenhouse Effect

Carbon dioxide (CO_2) is approximately 83% of the greenhouse gases emitted by the U.S. as a percent of the mass of carbon or carbon equivalent. T.M.L. Wigley, R. Richels and J.A. Edmonds [1996] projected ambient CO_2 concentration through the 21st century. Pre-industrial atmospheric CO_2 concentration was on the order of 288 parts per million. Atmospheric CO_2 concentration is currently at 340 parts per million. The concentration of CO_2 that would establish an acceptable energy balance is considered to be 550 parts per million. To achieve the acceptable concentration of CO_2 through the next century, society would have to reduce the volume of greenhouse gases entering the atmosphere.

The Kyoto Protocol is an international treaty that was negotiated in Kyoto, Japan in 1997 to establish limits on the amount of greenhouse gases a country can emit into the atmosphere. The Kyoto Protocol has not been accepted worldwide. Some countries believe the greenhouse gas emission limits are too low and would adversely impact national and world economies without solving the problem of global warming. Another criticism of the Kyoto Protocol is that it does not apply to all nations. For example, China is exempt from greenhouse gas emission limitations in the Kyoto Protocol even though it has one of the world's fastest growing economies and the world's largest population. Research is underway to develop the technology needed to capture and store greenhouse gases in geologic formations as an economically viable means of mitigating the increase in greenhouse gas concentration in the atmosphere.

Point to Ponder: Is geologic sequestration a desirable policy?

It would seem that geologic sequestration would be a desirable policy for a society that is dependent on fossil fuels and is trying to minimize carbon dioxide emissions. On the other hand, geologic sequestration can adversely

affect a society's willingness to move its energy infrastructure away from fossil fuels and towards cleaner alternate
energy sources. From this perspective, geologic sequestration is an undesirable policy.

RENEWABLE ENERGY AND THE ENVIRONMENT

The environmental impact of some clean energy sources is
summarized below:

- Hydroelectric facilities, notably dams, can flood vast
  areas of land. The flooded areas can displace people and
  wildlife, and impact the ecosystems of adjacent areas
  with consequences that may be difficult to predict. Dams
  can change the composition of river water downstream of
  the dam and can deprive land areas of a supply of silt for
  agricultural purposes. A dam on a river can prevent the
  upstream migration of certain species of fish, such as
  salmon.

- Geothermal power plants can emit toxic gases such as
  hydrogen sulfide or greenhouse gases such as carbon
  dioxide. The produced water from a geothermal reservoir
  will contain dissolved solids that can form solid precipi-
  tates when the temperature and pressure of the
  produced water changes.

- Solar power plants are relatively inefficient and a solar
  power plant like the Solar Electric Generating Station
  (SEGS) in Southern California has a large footprint and
  may be visually offensive to some people (Figure 4-10).

Figure 4-10 also shows some broken mirrors, which illustrates the need for maintenance to repair or replace damaged mirrors.

- Wind farms can interfere with bird migration patterns and may be visually offensive.

Point to Ponder: Is the environmental impact of clean energy sources an important consideration?

Society is searching for environmentally compatible, reliable energy sources. One of the characteristics of an environmentally compatible energy source is its cleanliness.

A clean energy source emits negligible amounts of greenhouse gases or other pollutants. Even though fossil fuels have serious pollution problems, we have seen that mitigation technologies such as greenhouse gas storage (also known as sequestration) can reduce the impact of greenhouse gas emissions and justify the continued use of fossil fuels as an energy source.

In an objective assessment of competing energy sources, we must recognize that clean energy sources such as hydroelectric, solar, and wind energy can have a significant environmental impact that can adversely affect their environmental compatibility. This impact may affect the social acceptability of renewable energy projects. For example, a wind farm proposed for installation in Nantucket Sound near Martha's Vineyard encountered strong political resistance from residents and some environmentalists because of its expected impact on the aesthetics of the area and the local fishing industry.

NUCLEAR ENERGY AND THE ENVIRONMENT

The selection of an environmentally compatible primary fuel is not a trivial problem. Two energy sources of special concern for the 21st century energy mix are nuclear fission and nuclear fusion. We consider them in more detail below.

Nuclear Fission

Nuclear fission plants can produce more energy and operate continuously for longer periods of time than other power plants. Compared to fossil fuel driven power plants, nuclear fission plants require a relatively small mass of resource to fuel the nuclear plant for an extended period of time. Nuclear fission plants rely on a non-renewable resource: uranium-235. The Earth's inventory of uranium-235 will eventually be exhausted. Breeder reactors use the chain reaction that occurs in the reactor control rods to produce more fissionable material (specifically plutonium-239).

One of the main concerns of nuclear fission technology is to find a socially and environmentally acceptable means of disposing of fuel rods containing highly radioactive waste. The issue of waste is where most of the debate about nuclear energy is focused. The waste generated by nuclear fission plants emits biologically lethal radiation and can contaminate the site where it is stored for thousands of years. On the other hand, environmentally compatible disposal options are being developed. One disposal option is to store spent nuclear fuel in geologically stable environments.

Nuclear Fusion

The resources needed for the nuclear fusion reaction are abundant; they are the isotopes of hydrogen. The major component of fusion, deuterium, can be extracted from water, which is available around the world. Tritium, another hydrogen isotope, is readily available in lithium deposits that can be found on land and in seawater. Unlike fossil fuel driven power plants, nuclear fusion does not emit air pollution.

Nuclear fusion reactors are considered much safer than nuclear fission reactors. The amounts of deuterium and tritium used in the fusion reaction are so small that the instantaneous release of a large amount of energy during an accident is highly unlikely. The fusion reaction can be shut down in the event of a malfunction with relative ease. A small release of radioactivity in the form of neutrons produced by the fusion reaction may occur, but the danger level is much less than that of a fission reactor. The main problem with nuclear fusion energy is that the technology is still under development; commercially and technically viable nuclear fusion reactor technology does not yet exist. A panel working for the United States Department of Energy has suggested that nuclear fusion could be providing energy to produce electricity by the middle of the 21st century if adequate support is provided to develop nuclear fusion technology [Dawson, 2002].

Radioactive Waste
Nuclear fission power plants generate radioactive wastes that require long-term storage. The issue of the disposal of radioactive byproducts created by nuclear reactors and the effects of nuclear waste on the environment has hindered the expansion of nuclear power. The end products of nuclear fission are highly radioactive and have a half-life measured in thousands of years. They must be disposed of in a way that offers long-term security. One solution is to place used uranium rods containing plutonium and other dangerously radioactive compounds in water that speeds the decay process of these products. The rods are then buried in a remote location that should not be significantly effected by low levels of radiation. Yucca Mountain, Nevada (Figure 9-5) is the location that is currently being considered for long-term nuclear waste storage in the United States.

A nuclear power plant can contaminate air, water, the ground, and the biosphere. Air can be contaminated by the release of radioactive vapors and gases through water vapor from the cooling towers, gas and steam from the air ejectors, ventilation exhausts, and gases removed from systems having radioactive fluids and gases. The radiation released into the air can return to the Earth as radiated rain, which is the analog to acid rain generated by the burning of fossil fuels.

Figure 9-5. Yucca Mountain (left) and Train inside Yucca Mountain Tunnel (right)

Water may be contaminated when radioactive materials leak into coolant water. The contaminated water can damage the environment if it is released into nearby bodies of water such as streams or the ocean. Water and soil contamination can occur when radioactive waste leaks from storage containers and seeps into underground aquifers. The biosphere (people, plants, and animals) is affected by exposure to radioactive materials in the environment. The effect of exposure is cumulative and can cause the immune system of an organism to degrade.

Containment Failure

Radioactive materials can be released from nuclear reactors if there is a failure of the containment system. Containment can be achieved by the reactor vessel and by the containment dome enclosing the reactor vessel. The two most publicized nuclear power plant incidents were containment failures. They occurred at Three Mile Island, Pennsylvania in 1979 and at Chernobyl, Ukraine in 1986.

The Three Mile Island power plant was a pressurized water reactor that went into operation in 1978 and produced approximately a gigawatt of energy. The containment failure at Three Mile Island occurred on March 29, 1979. It began when coolant feedwater pumps stopped and temperature in the reactor vessel began to rise. An increase in pressure accompanied the increasing temperature and caused a pressure relief valve to open. The reactor shut down automatically. Steam from the reactor flowed through the open relief valve into the containment dome. The valve failed to shut at a pre-specified pressure and vaporized coolant water continued to flow out of the reactor vessel through the open valve. The water-steam mixture flowing through the coolant pumps caused the pumps to shake violently. Plant operators did not realize they were losing coolant and decided to shut off the shaking pumps. A large volume of steam formed in the reactor vessel and the overheated nuclear fuel melted the metal tubes holding the nuclear fuel pellets. The exposed pellets reacted with water and steam to form a hydrogen gas bubble. Some of the hydrogen escaped into the containment dome. The containment dome did not fail; it contained the hydrogen gas bubble and pressure fluctuations. The operators were eventually able to disperse the hydrogen bubble and regain control of the reactor.

The Chernobyl containment failure occurred in a boiling water reactor that produced a gigawatt of power and had a graphite moderator. Operators at the plant were using one of the reactors, Unit 4, to conduct an experiment. They were testing the ability of the plant to provide electrical power as the reactor was shut down. To obtain measurements, the plant operators turned off some safety systems, such as the emergency cooling system, in violation of safety rules. The operators then withdrew the reactor control rods and shut off the generator that provided power to the cooling water pumps. Without coolant, the reactor overheated. Steam explosions exposed the reactor core and fires started. The Chernobyl reactors were not encased in massive containment structures that are common elsewhere in the world. When the explosions exposed the core, radioactive materials were released into the environment and a pool of radioactive lava burned through the reactor floor. The Chernobyl accident was attributed to design flaws and human error.

Except for the Chernobyl incident, no deaths have been attributed to the operation of commercial nuclear reactors. M.W. Carbon [1997, Chapter 5] reported that the known death toll at Chernobyl was less than fifty people. R.A. Ristinen and J.J. Kraushaar [1999, Section 6.9] reported an estimate that approximately 47,000 people in Europe and Asia will die prematurely from cancer because they were exposed to radioactivity from Chernobyl. The disparity in the death toll associated with the Chernobyl incident illustrates the range of conclusions that can be drawn by different people with different perspectives.

Nuclear Fallout

Nuclear fallout is the deposition of radioactive dust and debris that was carried into the atmosphere by the detonation of a nuclear weapon. Nuclear fallout can also be generated by the emission of nuclear material from a nuclear reactor that has been exposed to the atmosphere. The point of deposition of the fallout depends on climatic conditions such as ambient air temperature and pressure, and wind conditions.

Nuclear Winter

One of the more controversial issues associated with the climatic effects of nuclear fallout is the concept of nuclear winter. Nuclear winter is the prediction of significant climatic temperature declines resulting from an increase of atmospheric particulates following the detonation of many nuclear weapons in the atmosphere. The temperature decline would generate wintry conditions, hence the phrase "nuclear winter."

In the nuclear winter scenario, the temperature of the atmosphere is expected to decrease because of increased reflection of solar radiation as a result of additional particulates in the atmosphere. The same type of phenomenon is thought to have occurred during extinction events associated with meteors striking the Earth. One example of such an extinction event was the disappearance of the dinosaurs approximately 65 million years ago. Opponents of the nuclear winter scenario argue that the greenhouse effect may tend to increase the surface temperature of the Earth. It is now believed that the temperature decline would not be as severe as originally thought.

Point to Ponder: Is nuclear power socially acceptable?
Nuclear power has been socially acceptable in some political jurisdictions around the world. Some European states, notably France, are reliant on nuclear fission power. The acceptability of nuclear power depends on such factors as finding an environmentally acceptable solution for the storage of nuclear wastes and the cost of energy using non-nuclear energy sources. As the price of non-nuclear energy to the consumer increases, nuclear energy may become more appealing. The environmental concern over the geologic storage of radioactive waste must be weighed against the environmental concern over the emission and accumulation of greenhouse gases in the atmosphere because of the continuing reliance on fossil fuels.

Chapter 10

ENERGY FORECASTS

As a global society, we must face the need to reduce our reliance on oil. The issue is not whether oil will be replaced, but when. Decisions are being made today by companies and governments around the world based on the assumption that oil will be replaced as the primary source of energy. Governments of oil consuming nations are maintaining military forces capable of keeping oil supply lines open between producers and consumers. Oil exporting nations are trying to optimize their revenues by influencing the market price of oil. How high will the price of oil go as supply dwindles? What will we use to replace oil in the future energy mix? Every one of us has a stake in the answers to these questions and the decisions that are being made to provide energy.

One of the problems facing society is the need to develop and implement a strategy that will provide energy to meet future global energy needs and satisfy environmental objectives. The development of strategies depends on our view of the future. At best, we can only make educated guesses about what the future will bring. The quality of our educated guesses depends on our understanding and the information we have available.

Figure 10-1 displays three different levels of fore-casting, or predicting, the future: stories, scenarios, and models. A story can be used to provide a qualitative picture of the future. Stories are relatively unclear because our understanding is limited and the information we have is relatively incomplete. As we gain information and under-standing, we can begin to discuss scenarios. Scenarios let us consider different stories about complex situations. They let us incorporate more detail into plausible stories. Unlike forecasts, which let us extrapolate historical behavior to predict the future, scenarios let us consider the effects of discontinuities and sudden changes. Forecasts assume a certain degree of continuity from past to future, while the future may in fact be altered dramatically by an unexpected development. In the context of energy, revolutionary devel-opments in nuclear fusion technology, unexpected cost reductions in solar energy technology, or significant changes in the current world order can lead to abrupt changes in historical trends of energy production and consumption. These changes could invalidate any forecast that was based on a continuous extrapolation of historical trends and lead to a future that would have been considered implausible based on past performance.

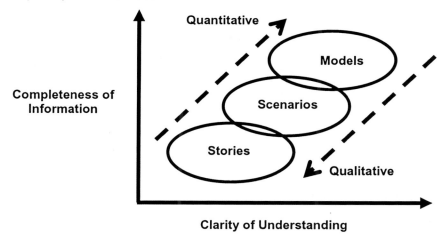

Figure 10-1. Stories, Scenarios and Models

We can construct models that allow us to quantify our scenarios as our knowledge and understanding increase. The concept of sustainable development is a road map of how we should prepare for the future and it is a vision of what the future should be. It is a scenario that has been adopted by the United Nations and helps explain evolving business practices in the energy industry. We discuss sustainable development as a road map for the future and present some forecasts that recognize the need to replace oil.

SUSTAINABLE DEVELOPMENT

An emerging energy mix is needed to meet energy demand in the 21st century. The demand for energy is driven by

factors such as increasing trends in population and consumption. The ability to meet the demand for energy
depends on such factors as price volatility, supply availability, and efficiency of energy use. One measure of how
efficiently a country is using its energy is energy intensity. In
the context of energy policy, energy intensity may be defined
on the national level as the total domestic primary energy
consumption divided by the gross domestic product. Countries that have low energy consumption and high domestic
productivity will have relatively low energy intensities. Countries with high energy consumption and low domestic
productivity will have relatively high energy intensities. By
considering the change in energy intensity as a function of
time, we can see if a country is improving its efficiency of
energy consumption relative to its domestic productivity.

Table 10-1 presents the gross domestic product (GDP
in billions of 1995 U.S. dollars) and primary energy consumption (in quadrillion BTUs) for a few countries for years
1990, 1995, and 2000. Data are from the U.S. Energy Information Administration website and were recorded in 2002.
The energy intensity (in barrels of oil equivalent per
US$1000) is shown in the last column [Fanchi, 2004, Exer-
cise 14-12]. The energy intensity is increasing as a function
of time for Saudi Arabia and is relatively constant for France
and Japan. The energy intensity is decreasing for China, the
United Kingdom, and the United States. If one of our goals is
to maintain or improve quality of life with improved energy

efficiency, we would like to see the energy intensity of a nation decrease as a function of time.

Table 10-1. Energy Intensity of Selected Countries				
Country	Year	GDP	Primary Energy Consumption	Energy Intensity
China	1990	398	27.0	11.7
	1995	701	35.2	8.64
	2000	1042	36.7	6.05
France	1990	1478	8.81	1.02
	1995	1555	9.54	1.06
	2000	1764	10.4	1.02
Japan	1990	4925	17.9	0.63
	1995	5292	20.8	0.68
	2000	5342	21.8	0.70
Saudi Arabia	1990	114	3.15	4.75
	1995	128	3.85	5.17
	2000	139	4.57	5.65
United Kingdom	1990	1041	9.29	1.53
	1995	1127	9.60	1.46
	2000	1295	9.88	1.31
United States	1990	6580	84.4	2.21
	1995	7400	91.0	2.11
	2000	9049	98.8	1.88

The emerging energy mix is expected to rely on clean energy, that is, energy that is generated with minimal environmental impact. The goal is sustainable development. The concept of sustainable development was introduced in 1987 in a report prepared by the United Nations' World Commission on Environment and Development. The Commission, known as the Brundtland Commission after chairwoman Gro Harlem Brundtland of Norway, said that society should adopt a policy of sustainable development that allows society to meet its present needs while preserving the ability of future generations to meet their own needs [WCED, 1987].

Society desires, and industry is seeking to achieve sustainable development. One industry response to environmental and social concerns in the context of sustainable development is the 'triple bottom line' [Whittaker, 1999]. The three components of sustainable development and the three goals of the triple bottom line (TBL) are economic prosperity, social equity, and environmental protection. From a business perspective, the focus of TBL is the creation of long-term shareholder value by recognizing that corporations are dependent on licenses provided by society to do business. If business chooses not to comply with sustainable development policies, society can enforce compliance by imposing additional government regulation and control.

Sustainable development is a scenario that is concerned about the rights of future generations. The concept of rights is a legal and philosophical concept. It is possible to argue that future generations do not have any legal rights to current natural resources and are not entitled to rights. From this perspective, each generation must do the best it can with available resources. On the other hand, many societies are choosing to adopt the value of preserving natural resources for future generations. National parks are examples of natural resources that are being preserved.

Point to Ponder: How do we estimate the size of the population in the future?

Biological population models may be used to estimate the size of the population in the future. We can use an exponential growth model that extrapolates historical trends without recognizing resource limitations or use a more sophisticated model. One example of a more sophisticated model is the logistic growth model, which uses historical data and includes a parameter called the carrying capacity that lets us represent feedback mechanisms that constrain the growth of a biological population. [Fanchi, 2004, Exercises 13-5 and 13-6] The projected size of the global population is an important factor in energy forecasts.

ENERGY AND ETHICS

One issue that must be considered in the context of sustainable development is the distribution of energy. Should energy be distributed around the world based on need, ability to pay, or some other value? This question is an ethical issue because the answer to the question depends on the values we choose to adopt.

The distribution of energy in the future will depend on whether or not a nation has a large per capita energy base. Should nations with energy resources help those in need? If so, how should they help? Traditional ethics would favor a policy of helping those nations without energy resources, but opinions differ on how to proceed. Two of the more important ethical positions are identified as *lifeboat ethics* and *spaceship ethics*. These positions are considered here for two reasons: they are diametrically opposed ethical positions that apply to the global distribution of energy; and they illustrate that people of good will can take opposite positions in a significant debate.

Proponents of *lifeboat ethics* oppose the transfer of wealth by charitable means. In this view, the more developed industrial nations are considered rich boats and the less developed, overcrowded nations are poor boats. The rich boats should not give the poor boats energy because their help would discourage the poor boats from making difficult choices such as population control and investment in infrastructure. Lifeboat ethics is a "tough love" position; it encourages nations to seek self-sufficiency. On the other hand, it might make some nations desperate and encourage the acquisition of energy resources by military means.

Proponents of *spaceship ethics* argue that everyone is a passenger on spaceship Earth. In this view, some passengers travel in first class while others are in steerage.

A more equitable distribution of energy is needed because it is morally just and it will prevent revolts and social turmoil. Thus, the wealthy should transfer part of their resources to the poor for both moral and practical reasons. On the other hand, nations that receive charitable donations of energy may be unwilling to make the sacrifices needed to become self-sufficient.

It may be that a synthesis of these two ethical positions would be the optimum policy for distributing energy. Countries with resources could help other countries develop their resources with the clear understanding that the developing countries must become self-sufficient. This level of cooperation between nations would be something new for the world to witness.

Point to Ponder: Do we need to consider the ethics of energy distribution?

The issue of energy distribution is really an issue of energy access. If people do not have access to essential resources, they may feel compelled to fight for those resources. Ethical positions such as *lifeboat ethics* and *spaceship ethics* attempt to address these concerns. Can access to energy be provided without discouraging self-reliance? Can the amount of useful energy be increased so that the need for sacrifice is lessened and the cost of energy is decreased? These are questions facing us today.

ENERGY AND GEOPOLITICS

Quality of life, energy, and the distribution of energy are important components of global politics. Readily available, reasonably priced energy is a critical contributor to the economic well-being of a nation. We have already seen that deforestation in England motivated the search for a new primary fuel. The need for oil encouraged Japanese expansion throughout Asia in the 1930's and was one of the causes of World War II. The 1973 Arab-Israeli war led to the first oil crisis with a short-term, but significant increase in the price of oil. This oil price shock was followed by another in 1979 after the fall of the Shah of Iran. These oil price increases are considered shocks because they were large enough to cause a significant decline in global economic activity [Verleger, 2000, page 76]. Our ability to correctly forecast energy demand depends on our understanding of technical and socio-political issues. In this section, we give a brief introduction to global politics and then discuss its implications for models of future energy demand.

Clash of Civilizations

The world has been undergoing a socio-political transition that began with the end of the Cold War and is continuing today. S.P. Huntington [1996] provided a view of this transition that helps clarify historical and current events, and provides a foundation for understanding the socio-political issues that will affect energy demand.

Huntington argued that a paradigm shift was occurring in the geopolitical arena. A paradigm is a model that is realistic enough to help us make predictions and understand events, but not so realistic that it tends to confuse rather than clarify issues. A paradigm shift is a change in paradigm. We can better appreciate the importance of a paradigm shift in geopolitics by first considering the value of geopolitical models.

A geopolitical model has several purposes. It lets us order events and make general statements about reality. We can use the model to help us understand causal relationships between events and communities. The communities can range in size from organizations to alliances of nations. The geopolitical model lets us anticipate future developments and, in some instances, make predictions. It helps us establish the importance of information in relation to the model and it shows us paths that might help us reach our goals.

The Cold War between the Soviet Union and the Western alliance led by the United States established a framework that allowed people to better understand the relationships between nations following the end of World War II in 1945. When the Cold War ended with the fall of the Berlin wall and the break up of the Soviet Union in the late 1980's and 1990's, it signaled the end of one paradigm and the need for a new paradigm. Several geopolitical models have been proposed. Huntington considered four possible paradigms for understanding the transition (Table 10-2).

Table 10-2. Huntington's Possible Geopolitical Paradigms	
1	One Unified World
2	Two Worlds (West versus non-West; us versus them)
3	Anarchy (184+ Nation-states)
4	Chaos

The paradigms in Table 10-2 cover a wide range of possible geopolitical models. The One Unified World paradigm asserts that the end of the Cold War signaled the end of major conflicts and the beginning of a period of relative calm and stability. The Two Worlds paradigm views the world in a "us versus them" framework. The world was no longer divided by political ideology (democracy versus communism); it was divided by some other issue. Possible divisive issues include religion and rich versus poor (generally a North-South geographic division). The world could also be split into zones of peace and zones of turmoil. The third paradigm, Anarchy, views the world in terms of the interests of each nation, and considers the relationships between nations to be unconstrained. These three paradigms, One Unified World, Two Worlds, and Anarchy, range from too simple (One Unified World) to too complex (Anarchy).

The Chaos paradigm says that post-Cold War nations are losing their relevance as new loyalties emerge. In a world where information flows freely and quickly, people are forming allegiances based on shared traditions and value systems. The value systems are notably cultural and, on a more fundamental level, religious. The new allegiances are in many cases a rebirth of historical loyalties. New alliances are forming from the new allegiances and emerging as a small set of civilizations. The emerging civilizations are characterized by ancestry, language, religion, and way of life.

Huntington considered the fourth paradigm, Chaos, to be the most accurate picture of current events and recent trends. He argued that the politics of the modern world can be best understood in terms of a model that considers relationships between the major contemporary civilizations shown in Table 10-3. The existence of a distinct African civilization has been proposed by some scholars, but is not as widely accepted as the civilizations identified in the table.

Each major civilization has at least one core state [Huntington, 1996, Chapter 7]. France and Germany are core states in the European Union. The United States is a core state in Western Civilization. Russia and China are core states, perhaps the only core states, in Orthodox Civilization and Sinic Civilization respectively. Core states are sources of order within their civilizations. Stable relations between core states can help provide order between civilizations. If energy resources are not equally distributed amongst civilizations, which they are not, then energy can become a significant factor in defining the relationships between civilizations.

Table 10-3. Huntington's Major Contemporary Civilizations	
Civilization	Comments
Sinic	China and related cultures in Southeast Asia
Japanese	The distinct civilization that emerged from the Chinese civilization between 100 and 400 C.E.
Hindu	The peoples of the Indian subcontinent that share a Hindu heritage.
Islamic	A civilization that originated in the Arabian peninsula and now includes subcultures in Arabia, Turkey, Persia, and Malaysia.
Western	A civilization centered around the northern Atlantic that has a European heritage and includes peoples in Europe, North America, Australia, and New Zealand.
Orthodox	A civilization centered in Russia and distinguished from Western Civilization by its cultural heritage, including limited exposure to Western experiences (such as the Renaissance, the Reformation, and the Enlightenment).
Latin America	Peoples with a European and Roman Catholic heritage who have lived in authoritarian cultures in Mexico, Central America and South America.

FORECASTS

Forecasts of energy production depend on the ability of energy producers to have access to natural resources.

Access depends, in turn, on the nature of relationships between civilizations with the technology to develop natural resources and civilizations with territorial jurisdiction over the natural resources. Examples of forecasts are discussed below.

Forecasts: Nuclear Energy
P.E. Hodgson [1999] presented a scenario in which the world would come to rely on nuclear fission energy. He defined five Objective Criteria for evaluating each type of energy: capacity, cost, safety, reliability, and effect on the environment. The capacity criterion considered the ability of the energy source to meet future energy needs. The cost criterion considered all costs associated with an energy source. The safety criterion examined all safety factors involved in the practical application of an energy source. This includes hazards associated with manufacturing and operations. The reliability criterion considered the availability of an energy source. By applying the five Objective Criteria, Hodgson concluded that nuclear fission energy was the most viable technology for providing global energy in the future.

According to Hodgson, nuclear fission energy is a proven technology that does not emit significant amounts of greenhouse gases. He argued that nuclear fission reactors have an exemplary safety record when compared in detail with other energy sources. Breeder reactors could provide the fuel needed by nuclear fission power plants and nuclear waste could be stored in geological traps. The security of nuclear power plants in countries around the world would be assured by an international agency such as the United Nations. In this nuclear scenario, renewable energy sources would be used to supplement fission power and fossil energy use would be minimized. Hodgson did not assume that the problems associated with nuclear fusion would be overcome. If they are, nuclear fusion could also be incorporated into the energy mix.

Forecasts: Renewable Energy

H. Geller [2003] presented a renewable energy scenario that sought to replace both nuclear energy and fossil energy with renewable energy only. An important objective of his fore-cast was to reduce greenhouse gas emissions to levels that are considered safe by the Kyoto Protocol. Figure 10-2 summarizes Geller's energy forecast. The figure shows global energy use as a function of time. Global energy use is expressed in billions of tons of oil equivalent (BTOE).

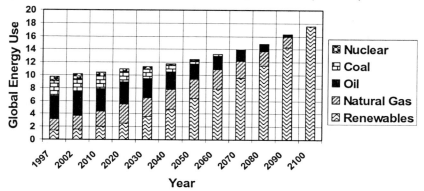

Figure 10-2. Forecast of a 21st Century Energy Mix that Eventually Relies on Renewable Energy Only

Forecasts: Schollnberger Energy Mix

Forecasts of the 21st century energy mix show that a range of scenarios is possible. The forecast discussed here is based on W.E. Schollnberger's forecasts, which were designed to predict the contribution of a variety of energy sources to the 21st century energy portfolio. Schollnberger's forecast is worth studying because it uses more than one scenario to project energy consumption for the entire 21st century.

Schollnberger considered the following three forecast scenarios:

A. "Another Century of Oil and Gas" corresponding to continued high hydrocarbon demand;

B. "The End of the Internal Combustion Engine" corresponding to a low hydrocarbon demand scenario; and

C. "Energy Mix" corresponding to a scenario with intermediate demand for hydrocarbons and an increasing demand for alternative energy sources.

Schollnberger viewed Scenario C as the most likely scenario. It is consistent with the observation that the transition from one energy source to another has historically taken several generations. Leaders of the international energy industry have expressed a similar view that the energy mix is undergoing a shift from liquid fossil fuels to other fuel sources.

There are circumstances in which Scenarios A and B could be more likely than Scenario C. For example, Scenario B would be more likely if environmental issues led to political restrictions on the use of hydrocarbons and an increased reliance on conservation. Scenario B would also be more likely if the development of a commercially competitive fuel cell for powering vehicles reduced the demand for hydrocarbons as a transportation fuel source. Failure to develop alternative technologies would make Scenario A more likely. It assumes that enough hydrocarbons will be supplied to meet demand.

Scenario C shows that natural gas will gain in importance as the economy shifts from a reliance on hydrocarbon liquid to a reliance on hydrocarbon gas. Eventually, renewable energy sources such as biomass and solar will displace oil and gas (see Figure 10-3). The vertical axis in Figure 10-3 presents energy in quads.

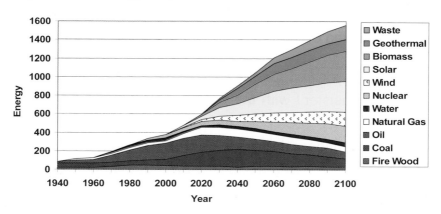

Figure 10-3. Forecast of 21st Century Energy Consumption

The demand by society for petroleum fuels should continue at or above current levels for a number of years, but the trend seems clear (see Figure 10-4). The global energy portfolio is undergoing a transition from an energy portfolio dominated by fossil fuels to an energy portfolio that includes a range of fuel types. Schollnberger's Scenario C presents one possible energy portfolio and the historical and projected energy consumption trends are illustrated in Figure 10-4. The vertical axis in Figure 10-4 presents energy consumption as a percentage of energy consumed.

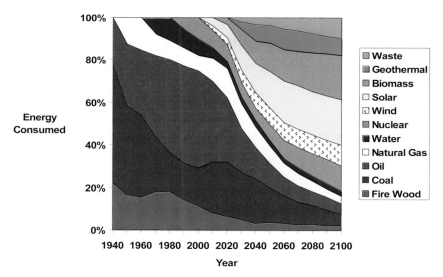

Figure 10-4. Forecast of 21st Century Energy Consumption as % of Energy Consumed

Point to Ponder: What role will energy conservation play in the future energy mix?
Energy can be conserved using a number of very simple

techniques. Some simple energy conservation methods that each of us can adopt include walking more and driving less; carpooling; planning a route that will let you drive to all of your destinations in the shortest distance; and using passive solar energy to dry your clothes or heat your home. Other energy conservation methods are more complicated: they are designed to reduce energy losses by improving the energy conversion efficiency of energy consuming devices. For example, improving gas mileage by reducing the weight of a vehicle or increasing the efficiency of internal combustion engines can reduce energy loss. We can decrease the demand for energy consuming activities such as air conditioning in the summer or heating in the winter by developing and using more effective insulating materials.

Energy conservation may be improved by increasing the efficiency of converting energy from one form into another. This efficiency is called energy conversion efficiency. It is the ratio of energy output to energy input. If we can decrease energy lost or wasted by a system, we can increase energy conversion efficiency. One method for improving energy conversion efficiency is to find a way to use energy that would otherwise be lost as heat.

Cogeneration is the simultaneous production and application of two or more sources of energy. The most common example of cogeneration is the simultaneous generation of electricity and useful heat. In this case, a fuel like natural gas can be burned in a boiler to produce steam. The steam drives an electric generator and is recaptured for such purposes as heating or manufacturing. Cogeneration is most effective when the cogeneration facility is near the site where excess heat can be used. The primary objective of cogeneration is to reduce the loss of energy by converting part of the energy loss to an energy output.

The social acceptability of energy conservation varies widely around the world. In some countries such as Germany, energy conservationists and environmentalists are a political force (the Green Party). In other countries, people may espouse conservation measures but be unwilling to participate in or pay for energy conservation practices, such as recycling or driving energy efficient vehicles. Some governments, especially in energy importing nations, are encouraging or requiring the development of energy conserving technologies. We can expect energy conservation to increase in the future as a result of increasing energy costs, more widespread adoption of energy conservation measures, and by improvements in energy conversion efficiency. We should not, however, believe that energy conservation will be enough to satisfy global energy needs.

Forecasts: Hubbert and a Gaussian Fit

Schollnberger's forecast is based on demand. An alternative approach is to base the forecast on supply. Beginning with

M.K. Hubbert [1956], several authors have noted that annual U.S. and world oil production approximately follows a bell shaped (Gaussian) curve.

Forecasts based on Gaussian fits to historical data can be readily checked using publicly available data. Figure 10-5 shows a Gaussian curve fit of world oil production data from the U.S. EIA database. The fit is designed to match the most recent part of the production curve most accurately. This gives a match that is similar to results obtained by K.S. Deffeyes [2001, page 147]. The peak oil production rate in Figure 10-5 below occurs in 2010 and cumulative oil production by year 2100 is a little less than 2.1 trillion barrels.

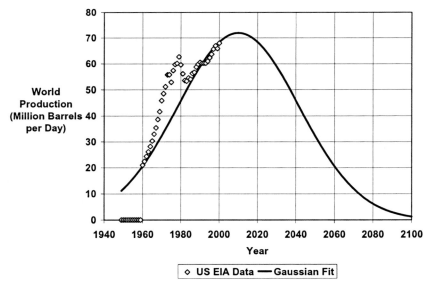

Figure 10-5. Oil Forecast Using Gaussian Curve

Analyses of historical data using a Gaussian curve typically predict that world oil production will peak in the first decade of the 21st century. By integrating the area under the Gaussian curve, forecasters have claimed that cumulative world oil production will range from 1.8 to 2.1 trillion barrels. These forecasts usually underestimate the sensitivity of oil production to technical advances and price. In addition, forecasts often discount the large volume of oil that has been discovered but not yet produced because the cost of production has been too high.

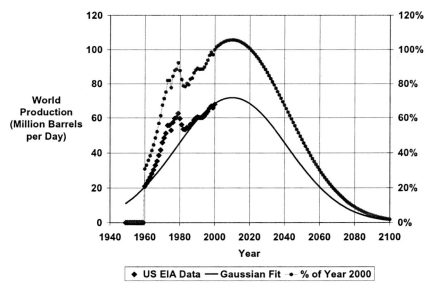

Figure 10-6. Oil Forecast as % of Oil Produced in Year 2000 Using Gaussian Curve

If we accept a Gaussian fit of historical data as a reasonable method for projecting oil production, we can estimate future oil production rate as a percentage of oil production rate in the year 2000. Figure 10-6 shows this estimate. According to this approach, world oil production rate will decline to 50% of year 2000 world oil production by the middle of the 21st century. For comparison, let us consider Schollnberger's Scenario C.

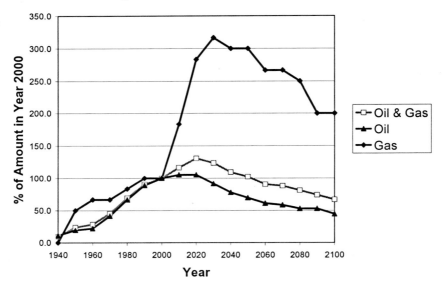

Figure 10-7. Forecast of 21st Century Oil and Gas Consumption as % of Oil and Gas Consumed in Year 2000

According to Scenario C, fossil fuel consumption will increase relative to its use today until about the middle of the 21st century, when it will begin to decline (see Figure 10-7). By the end of the 21st century, fossil fuel consumption will be approximately 70% of what it is today. Gas consumption will be considerably larger, while oil consumption will decline to approximately half of its use today. A comparison of oil production as a percent of year 2000 production in Figure 10-6 with oil consumption as percent of oil consumed in year 2000 illustrates the range of uncertainty in existing forecasts.

The supply-based forecast shows that oil production will approach 0% of year 2000 production by year 2100. By contrast, the demand-based forecast shown in Figure 10-7 expects oil to be consumed at about 50% of its year 2000 consumption. There is a clear contradiction between the two forecasts that can be used to test the validity of the forecasts. Another test of forecast validity is the peak of world oil production.

Forecasts of world oil production peak tend to shift as more historical data is accumulated. J.H. Laherrère [2000] pointed out that curve fits of historical data should be applied to activity that is "unaffected by political or significant economic interference, to areas having a large number of fields, and to areas of unfettered activity" (pg. 75). Furthermore, curve fit forecasts work best when the inflection point (or peak) has been passed. Given these provisions, we see from Figure 10-5 that an oil peak will occur between 2000 and 2020.

Another way to view the peak in world oil production is to ask when the peak oil production will occur per person. This view recognizes that oil is a source of energy for a growing global population. Figure 10-8 shows a plot of per capita oil production (in gallons of oil per day per person) and the size of the global population (in millions of people). From this perspective, the maximum production of oil per person occurred in 1979, and was about 0.6 gallons of oil per day per person.

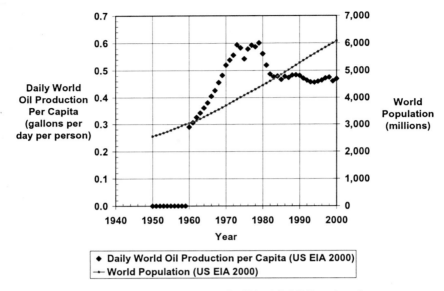

Figure 10-8. Historical Daily World Oil Production
per Capita and World Population

Point to Ponder: Which energy mix forecast is most likely?

The validity of an energy mix forecast depends on technical feasibility, economic viability, and political decisions. Social concern about nuclear waste and proliferation of nuclear weapons is a significant deterrent to reliance on nuclear fission power. Society's inability to resolve the issues associated with nuclear fusion make fusion an unlikely contributor to the energy mix until the middle of the 21st century. Concerns about greenhouse gas emissions and global warming are encouraging a movement away from fossil fuels. Political instability in countries that export oil and gas, and the finite size of known oil and gas supplies are contributing to the adoption of alternative energy sources. Increases in fossil fuel prices and decreases in renewable energy costs, especially wind energy, are encouraging the adoption of renewable energy. [Fanchi, 2004, Exercises 15-4 through 15-8]

The combined impact of increased fossil fuel costs and political instability may accelerate the decline in consumption of fossil fuels and an increase in reliance on renewable sources. We cannot say with certainty what the 21st century energy mix will be. We can say that the 21st century energy mix will depend as much on choices made by society as it will on technological advances.

Figure Credits

Figure 1-1. Fanchi, 2000
Figure 1-2. Fanchi, 2003
Figure 1-5. Fanchi, 2003
Figure 1-9. After Ausubel [2000]

Figure 2-4. Fanchi, 2002
Figure 2-6. Fanchi, 2003
Figure 2-7. Fanchi, 2004

Figure 3-1. Fanchi, 2004
Figure 3-2. After Cassedy and Grossman, 1998, page 177; and U.S. Department of Energy Report *Energy Technologies and the Environment*, Report Number DOE/EP0026
Figure 3-3. Fanchi, 2002

Figure 4-7. After Cassedy and Grossman, 1998, page 282; and U.S. Department of Energy, 1981

Figure 4-8. After Kraushaar and Ristinen, 1993, page 172; and Solar Energy Research Institute (now National Renewable Energy Laboratory), Golden, Colorado
Figure 4-10. Fanchi, 2002
Figure 4-11. Fanchi, 2002

Figure 5-2. Fanchi, 2003
Figure 5-5. Fanchi, 2002
Figure 5-6. Fanchi, 2003
Figure 5-7. After Shepherd and Shepherd, 1998, page 227; and J.C. McVeigh, *Energy Around the World*, Pergamon Press, Oxford, United Kingdom
Figure 5-8. After Shepherd and Shepherd, 1998, page 227; and *Renewable Energy – A Resource for Key Stages 3 and 4 of the UK National Curriculum*, Renewable Energy Enquires Bureau, Oxfordshire, United Kingdom
Figure 5-9. Fanchi, 2002
Figure 5-10. Fanchi, 2004
Figure 5-11. After Shepherd and Shepherd, 1998, page 149; and *Renewable Energy – A Resource for Key Stages 3 and 4 of the UK National Curriculum*, Renewable Energy Enquires Bureau, Oxfordshire, United Kingdom
Figure 5-12. Fanchi, 2004

Figure 6-1. Fanchi, 2003
Figure 6-2. Fanchi, 2002
Figure 6-3. Fanchi, 2003
Figure 6-4. Fanchi, 2004

Figure 6-5. After Cassedy and Grossman, 1998, page 298; and P.D. Dunn, *Renewable Energies: Sources, Conversion and Application*, 1986, P. Peregrinius, Ltd., London, United Kingdom

Figure 7-1. Fanchi, 2000
Figure 7-3. After Cassedy and Grossman, 1998, page 419; and B.J. Crowe, *Fuel Cells – A Survey*, 1973, NASA, U.S. Government Printing Office, Washington, D.C.
Figure 7-4. Fanchi, 2003

Figure 8-2. Fanchi, 2002
Figure 8-3. Fanchi, 2002
Figure 8-6. Fanchi, 2003

Figure 9-1: After Fanchi, 2001, Chapter 9
Figure 9-2. Fanchi, 2002
Figure 9-3. Fanchi, 2002
Figure 9-5. Courtesy Lander County, NV, 2004

Figure 10-1. After McKay, 2002
Figure 10-2. After Geller, 2003, page 227

References

Aubrecht, Gordon J., 1995, *Energy*, 2nd Edition, Prentice-Hall, Inc., Upper Saddle River, New Jersey.

Ausubel, J.H., 2000, "Where is Energy Going?" *The Industrial Physicist* (Feb.), pages 16-19.

Bain, A., and Van Vorst, W.D., 1999, "The Hindenburg tragedy revisited: the fatal flaw found," *International Journal of Hydrogen Energy*, Volume 24, pages 399-403.

Borbely, A. and Kreider, J.F., 2001, *Distributed Generation: The Power Paradigm for the New Millenium*, CRC Press, New York, New York.

Brennan, T.J., Palmer, K.L., Kopp, R.J., Krupnick, A.J., Stagliano, V. and Burtraw, D., 1996, *A Shock to the System: Restructuring America's Electricity Industry*, Resources for the Future, Washington, D.C.

Carbon, M.W., 1997, *Nuclear Power: Villain or Victim?*, Pebble Beach Publishers, Madison, Wisconsin.

Cassedy, E.S. and Grossman, P.Z., 1998, *Introduction to Energy*, 2ⁿᵈ Edition, Cambridge U.P., Cambridge, U.K..

Cook, E., 1971, "The Flow of Energy in an Industrial Society," *Scientific American* (Sep.), pages 135-144.

Dawson, J., 2002, "Fusion Energy Panel Urges US to Rejoin ITER," *Physics Today* (November), pages 28-29.

Deffeyes, K.S., 2001, *Hubbert's Peak – The Impending World Oil Shortage*, Princeton U.P., Princeton, New Jersey.

DoE Geothermal, 2002, "Geothermal Energy Basics," United States Department of Energy website (Accessed 23 October 2002)
http://www.eren.doe.gov/geothermal/geobasics.html.

DoE Hydropower, 2002, "Hydropower," United States Department of Energy website (Accessed 24 October 2002)
http://www.eren.doe.gov/RE/hydropower.html.

DoE Ocean, 2002, "Ocean," United States Department of Energy website (Accessed 24 October 2002)
http://www.eren.doe.gov/RE/Ocean.html.

EIA Table 6.2, 2002, "World Total Net Electricity Consumption, 1980-2000," United States Energy Information Administration website (Accessed 10 June 2002) http://www.eia.doe.gov/emeu/international/electric.html#IntlConsumption.

EIA Table 11.1, 2002, "World Primary Energy Production by Source," 1970-2000, United States Energy Information Administration website (Accessed 10 November 2002) http://www.eia.doe.gov/emeu/international/electric.html#IntlProduction.

EIA Table E.1, 2002, "World Primary Energy Consumption (Btu), 1980-2000," United States Energy Information Administration website (Accessed 10 June 2002) http://www.eia.doe.gov/emeu/international/total.html#IntlConsumption.

Fanchi, J.R., 2004, *Energy: Technology and Directions for the Future*, Elsevier-Academic Press, Boston.

Geller, H., 2003, *Energy Revolution*, Island Press, Washington.

Gold, Thomas, 1999, *The Deep Hot Biosphere*, Springer-Verlag New York, Inc., New York, New York.

Goswami, D.Y., Kreith, F., and Kreider, J.F., 2000, *Principles of Solar Engineering*, George H. Buchanan Co., Philadelphia, Pennsylvania.

Hayden, H.C., 2001, *The Solar Fraud: Why Solar Energy Won't Run the World*, Vales Lake Publishing, LLC, Pueblo West, Colorado.

Hodgson, P.E., 1999, *Nuclear Power, Energy and the Environment*, Imperial College Press, London, United Kingdom.

Hubbert, M.K., 1956, "Nuclear Energy and the Fossil Fuels," American Petroleum Institute Drilling and Production Practice, Proceedings of the Spring Meeting, San Antonio, pages 7-25.

Huntington, S.P., 1996, *The Clash of Civilizations*, Simon and Schuster, London.

Kraushaar, J.J. and Ristinen, R.A., 1993, *Energy and Problems of a Technical Society*, 2ⁿᵈ Edition, Wiley, New York.

Laherrère, J.H., 2000, "Learn strengths, weaknesses to understand Hubbert curves," *Oil and Gas Journal*, pages 63-76 (17 Apr.); see also Laherrère's earlier article "World oil supply – what goes up must come down, but when will it peak?" *Oil and Gas Journal*, pages 57-64 (1 Feb. 1999) and letters in *Oil and Gas Journal*, (1 Mar. 1999).

Lide, D.R., 2002, *CRC Handbook of Chemistry and Physics*, 83rd Edition, CRC Press, Boca Raton, Florida.

Lilley, J., 2001, *Nuclear Physics*, Wiley, New York.

McKay, D.R. (correspondent), 2002, "Global Scenarios 1998-2020," Summary Brochure, Shell International, London.

Morrison, P. and Tsipis, K., 1998, *Reason Enough to Hope*, The MIT Press, Cambridge, Massachusetts, especially Chapter 9.

Murray, R.L., 2001, *Nuclear Energy: An Introduction to the Concepts, Systems, and Applications of Nuclear Processes*, 5th Edition, Butterworth-Heinemann, Boston, Massachusetts.

Nef, J.U., 1977, "An Early Energy Crisis and its Consequences," *Scientific American* (November), pages 140-151.

Ogden, J.M., 2002, "Hydrogen: The Fuel of the Future?" *Physics Today* (April), pages 69-75.

Ramage, J. and Scurlock, J., 1996, "Biomass," *Renewable Energy: Power for a Sustainable Future*, edited by G. Boyle, Oxford University Press, Oxford, United Kingdom.
Ristinen, R.A. and Kraushaar, J.J., 1999, *Energy and the Environment*, Wiley, New York.

Schollnberger, W.E., 1999, "Projection of the World's Hydrocarbon Resources and Reserve Depletion in the 21st Century," *The Leading Edge* (May 1999), pages 622-625.

Serway, R.A. and Faughn, J.S., 1985, *College Physics*, Saunders, Philadelphia.

Shepherd, W. and D.W. Shepherd, 1998, *Energy Studies*, Imperial College Press, London, U.K.

Silberberg, M., 1996, *Chemistry*, Mosby, St. Louis.

Sørensen, Bent, 2000, *Renewable Energy: Its physics, engineering, environmental impacts, economics & planning*, 2nd Edition, Academic Press, London, U.K.

UNDP, 2001, *Human Development Report 2001: Making New Technologies Work for Human Development*, United Nations Development Program, Oxford University Press, New York.

van Dyke, K., 1997, *Fundamentals of Petroleum*, 4th Edition, Petroleum Extension Service, University of Texas, Austin.

Verleger, P.K. Jr., 2000, "Third Oil Shock: Real or Imaginary?" *Oil & Gas Journal* (12 June), pages 76-88.

WCED (World Commission on Environment and Development), Brundtland, G., Chairwoman, 1987, *Our Common Future*, Oxford University Press.

Whittaker, M., 1999, "Emerging 'triple bottom line' model for industry weighs environmental, economic, and social considerations," *Oil and Gas Journal*, pages 23-28 (20 Dec.).

Wigley, T.M.L., Richels, R., and Edmonds, J.A., 1996, "Economic and environmental choices in the stabilization of atmospheric CO2 concentrations," *Nature* (18 Jan.), pages 240-243.

Wiser, W.H., 2000, *Energy Resources: Occurrence, Production, Conversion, Use*, Springer-Verlag New York, Inc., New York, New York.

Yergin, D., 1992, *The Prize*, Simon and Schuster, New York.

Index

cell 90, 98, 133, 137, 138, 141-143, 145, 146, 155, 214
chain reaction 54, 55, 62, 189
chaos 208, 209
Chernobyl 70, 193, 194
China 180, 186, 200, 201, 209, 210
chromosphere 73
civilization 13, 206, 209-211
cleat 34
closed system 107
coal 13, 25, 30-32, 34, 35, 150, 151, 156, 167
coal gasification 129, 132, 140
coal mine 34, 150, 151
coal seam 32, 33, 132
coalbed methane 32, 34, 42
coals 31
cogeneration 215
Cold War 206-208
combustion 7, 13, 17, 47, 120-124, 126, 127, 132, 133, 136, 141, 146, 184, 213, 215
composition 25, 31, 34-36, 187
conduction 83
conductor 90, 162, 164
conservation 80, 214, 215
containment failure 193, 194
control rod 189, 194
convection 73, 83
coolant 60, 82, 192-194
crust 29, 62, 108, 110

D
dam 18, 100-102, 105, 157, 158, 187
decarbonization 19, 20, 22, 143
deforestation 13, 47, 122, 151, 206
Denmark 46
desorption 35
direct current 152, 161

solar flux 75
solar heat collector 81, 83, 86
solar intensity 88
solar power 81, 84-86, 88, 90, 187
South Korea 68, 69
space 36, 75, 80, 108, 184
Spain 12, 69, 70
star 73, 74
Stefan-Boltzmann law 83
sunlight 74, 76, 77, 78, 79, 81, 84, 85, 86, 87, 90, 169, 184
superconductivity 133
sustainable development 144, 199, 202, 203
synfuels 10, 118, 128, 132

T
tar sand 43, 45, 182
temperature 25, 26, 31, 35, 37, 43, 73, 77, 79, 83, 108, 110-115, 162, 165, 183, 184
thermal conductivity 79, 80
Three Mile Island 70, 193
tidal energy 105, 106
tide 103, 105, 116
tight gas 42
time 14, 27, 30, 36, 37, 52, 74, 75, 100, 153, 157, 161
tokomak 66
transformer 159, 161, 163-165
transmission line 159-164
transmission system 163, 169
turbulence 96, 101
Turkey 210

U
Ukraine 58, 68-70, 193
United Arab Emirates 11
United Kingdom 11, 68, 69, 200, 201
United Nations 8-10, 18, 199, 202, 212

V

volcano 107, 109, 110

W

wave 103-106, 116
wave energy 104
well 14, 34-40, 43, 83, 112, 184
wind energy 45, 92, 93, 97, 98, 140, 168, 169, 188, 222
wind farm 96-98, 167, 188
wind park 96
wind turbine 93-98, 167
windmill 5, 6, 93, 149
wood 4-6, 13, 19, 118, 120-122, 129, 149, 151, 156, 168, 183
work 6, 5, 81, 89, 98, 104, 153, 159, 161, 168, 178, 221
work function 89

Y

Yucca 191, 192

OTHER BOOKS BY
JOHN R. FANCHI

ENERGY: Technology and Directions for the Future, Elsevier-Academic Press, Boston, MA (2004).

Shared Earth Modeling, Butterworth-Heinemann, Woburn, MA and Oxford, UK (2002).

Principles of Applied Reservoir Simulation, Second Edition, Butterworth-Heinemann, Woburn, MA and Oxford, UK (2001). Chinese edition published in 2001.

Math Refresher for Scientists and Engineers, Second Edition, J. Wiley - Interscience, New York (2000).

Integrated Flow Modeling, Elsevier, Amsterdam (2000).

Parametrized Relativistic Quantum Theory, Kluwer Academic Publishers, Dordrecht, The Netherlands (1993).

St. Louis Community College
at Meramec
LIBRARY